D0874469

STANDARD REFRIGERATION AND AIR CONDITIONING QUESTIONS AND ANSWERS

Plant Operators' Manual, STEPHEN M. ELONKA

Standard Boiler Operators' Questions and Answers, STEPHEN M. ELONKA AND ANTHONY L. KOHAN

Standard Electronics Questions and Answers, STEPHEN M. ELONKA AND JULIAN L. BERNSTEIN
Volume I, Basic Electronics
Volume II, Industrial Applications

Standard Industrial Hydraulics Questions and Answers, STEPHEN M. ELONKA AND ORVILLE H. JOHNSON

Standard Instrumentation Questions and Answers, STEPHEN M. ELONKA AND ALONZO R. PARSONS
Volume I, Measuring Systems
Volume II, Control Systems

Standard Plant Operator's Questions and Answers, STEPHEN M. ELONKA AND JOSEPH F. ROBINSON
Volume I
Volume II

Standard Refrigeration and Air Conditioning Questions and Answers, (2d ed.) STEPHEN M. ELONKA AND QUAID W. MINICH

STANDARD REFRIGERATION AND AIR CONDITIONING QUESTIONS AND ANSWERS

STEPHEN MICHAEL ELONKA

Contributing Editor, Power magazine; Licensed Chief Marine Steam Engineer, Oceans, Unlimited Horsepower; Licensed as Regular Instructor of Vocational High School, New York State; Member, National Association of Power Engineers (life, honorary); Author: The Marmaduke Surfaceblow Story, Plant Operators' Manual; Co-author: Standard Plant Operator's Questions and Answers, Volumes I and II; Standard Refrigeration and Air Conditioning Questions and Answers; Standard Instrumentation Questions and Answers, Volumes I and II; Standard Electronics Questions and Answers, Volumes I and II; Standard Boiler Operators' Questions and Answers; Standard Industrial Hydraulics Questions and Answers; Electrical Systems and Equipment for Industry; Plant Energy Systems; Mechanical Packing Handbook.

QUAID WALTON MINICH

Vice President, Engineering, Del E. Webb Realty & Management Co.; American Society of Heating, Refrigerating and Air-Conditioning Engineers; Gold Seal Refrigeration License, State of New Jersey

Second Edition

LIBRARY
ROLLING PLAINS CAMPUS
TEXAS STATE TECHNICAL INSTITUTE

McGRAW-HILL BOOK COMPANY

New York   St. Louis   San Francisco   Düsseldorf   Johannesburg
Kuala Lumpur   London   Mexico   Montreal   New Delhi
Panama   Rio de Janeiro   Singapore   Sydney   Toronto

Library of Congress Cataloging in Publication Data

Elonka, Stephen Michael, date
    Standard refrigeration and air conditioning questions
and answers.

    1. Questions and answers—Refrigeration and refri-
gerating machinery. 2. Questions and answers—Air
conditioning. I. Minich, Quaid Walton, joint author.
II. Title.
TP492.E4   1973   621.5'6'076   72-13017
ISBN 0-07-019291-X

Copyright © 1973, 1961 by McGraw-Hill, Inc. All rights reserved.
Printed in the United States of America. No part of this
publication may be reproduced, stored in a retrieval system, or
transmitted, in any form or by any means, electronic, mechanical,
photocopying, recording, or otherwise, without the prior written
permission of the publisher.
1 2 3 4 5 6 7 8 9   BPBP   7 6 5 4 3

*The editors for this book were Tyler G. Hicks and
Stanley E. Redka, and its production was supervised by George E. Oechsner.
It was set in Electra by Monotype Composition Company, Inc.
It was printed and bound by The Book Press.*

TP
492
.E4
1973

Dedicated to
today's service man and operating engineer of refrigeration, air
conditioning and cryogenic equipment, who must keep up with this
constantly expanding field if he is to advance in pay and
keep the ever more complex machinery and
instruments in good working order

27082

PREFACE

This second edition contains many major advances in design and operating practice since the 1961 printing. In addition to the new rotary screw compressor, we cover today's improvements in refrigeration systems, the latest techniques in troubleshooting, computer control of air-conditioning systems, and the current operators' license requirements in the U.S. and Canada, the only survey of these data published anywhere.

This is one of several books in the Standard Question and Answer library. It is the result of numerous letters written to *Power* magazine asking for information in one convenient source. The book should be helpful to:

1. Plant and building operators and other practical men who want to pass examinations in refrigeration and air conditioning and improve their basic knowledge.

2. Industrial and other firms with a training program for upgrading employees.

3. License examining boards (refrigeration, air conditioning, stationary engineers, building operators, etc.), as a source of questions in one reliable, up-to-date work to help them *standardize* their examinations and requirements. (A quick look at Chap. 22 will show that no two requirements are the same.)

4. Schools, especially short-course cram schools and trade schools, that seek a practical textbook. It is also intended as a text for preliminary courses in refrigeration and air conditioning, especially at the technical institute level.

5. Executives and personnel managers, to use as a guide when interviewing candidates for employment or upgrading.

6. Sales and service personnel, to gain a better background of plant equipment.

7. Architects and professional and consulting engineers, for reference in many practical areas (such as license requirements, Chap. 22) not usually found in handbooks.

8. Building maintenance men, machinery operators, stationary and other engineers, as a study reference for upgrading or for obtaining an original license.

The field of refrigeration, air conditioning, and cryogenics is expanding rapidly. Almost every type of plant and building erected today —as well as the better domestic homes—has such equipment. This book, and other books in the Standard Question and Answer library, contains a comprehensive sampling of what readers should know about up-to-date equipment and safety. If these books are widely used for testing and for study purposes, an operator passing one examination should be able to pass a similar examination in any part of the country.

The suggested references at the end of each chapter are important. Try to study as many of these references as possible; they will give you details that cannot be covered fully in any one book.

The Standard Question and Answer library is a significant step in training plant and building operators to understand today's modern equipment. It should help to elevate operating engineers to a position of higher pay and respect. Some of these volumes have been translated and published in foreign countries—Japan, Poland, and now India.

The authors are indebted to the many equipment manufacturers who cooperated wholeheartedly in supplying data and illustrations. The various boards of examiners for stationary, refrigeration, and air-conditioning engineers were a big help in indicating the areas in which they examine.

Deserving warm thanks is Wesley D. Turner, of Turner's Refrigeration School, James G. Manning, D. K. Chapman, William J. Wren, John G. Greenwood, L. C. Rinehart, Anthony L. Kohan and editors at Power magazine.

Stephen Michael Elonka
Quaid Walton Minich

CONTENTS

Preface vii

Chapter 1. FUNDAMENTALS 1

Heat • Temperature • Specific Heat • Sensible Heat • Latent Heat • Energy • Laws of Energy • Kinetic Theory of Heat • Pressure and Boiling Point • Absolute Pressure • Gage Pressure • Inches of Vacuum • Psia and Psig • Dalton's Law of Partial Pressures • Critical Pressure • Critical Temperature • Frigorific Mixture • Tons of Refrigeration • Heat-absorbing Capacity • Superheat • Glossary

Chapter 2. REFRIGERATION SYSTEMS 14

Early Refrigeration • Ice • Principle of Mechanical Refrigeration • Cooling Effect • Compression System • Compression Temperatures • Compression Pressures • Compression Ratio (CR) • Absorption System • Absorption System Efficiency • Analyzer or Bubble Column • Rectifier · Heat Exchangers · Low Side and High Side · Lithium Bromide Absorption System · Absorption versus Compression System · Steam-jet System • Booster System • Hermetic Centrifugal System

Chapter 3. REFRIGERANTS 27

Characteristics Desired—Water as a Refrigerant · Latent Heat of Evaporation · Density Effect on Capacity · Odorless Refrigerant · Critical Temperature · Boiling Point · Stability · Physical Properties · Vacuum Refrigerant · Halogenated Hydrocarbons · Freon · Numbering System · Halocarbons and Lube Oils · Ammonia and Lubricants · Moisture · Mixing Refrigerants · Replacing Refrigerants

Chapter 4. COMPRESSORS 37

Reciprocating Compressor • Centrifugal Compressor • Rotary Sliding-Vane Compressor • Helical Rotary Screw Compressor • HDA Compressor • VSA Compressor • V or VW Compressor • Angle Compressor • Booster Compressor • Hermetic Compressor • Y Compressor • Multiple-effect Compressor • Stuffing Box • Oil Lantern Ring • Mechanical Shaft Seal • Lubrication • Cylinder Water Jackets • Safety Head • Compressor Clearance • Clearance Pocket • Bypass Capacity Control • Reducing Com-

pressor Capacity · Unloader · Crossover Valves · Starting Bypass · Snifter
Valve · Rotary Screw Compressor

Chapter 5. CONDENSERS   55

Heat Handled by Condensers · Six Types of Condensers · Selecting a
Condenser · Air-cooled Condensers · Water per Ton for Cooling · Gallon-
degrees per Ton · Horizontal Shell-and-tube Condenser · Vertical Open
Shell-and-tube Condenser · Shell-and-coil Condenser · Double-pipe Water-
cooled Condenser · Atmospheric Condenser · Evaporative Condenser ·
Subcooling Coil · Precooling Coil · Desuperheating Coil · Make-up
Water · Capacity of Evaporative Condenser · Bleedoff Line · Subfreezing
Operation of Evaporative Condensers · Number of Systems Operated from
One Evaporative Condenser · Maintenance of Evaporative Condensers ·
Loss of Condensing Capacity · Foul Gases · Noncondensable Gases ·
Valves for Horizontal Shell-and-tube Condenser · Fouling Factor

Chapter 6. EVAPORATORS   66

Purpose of Evaporator · Two Main Types · Three Requirements in Design
· Prime Surface · Extended Surface · Advantages of Extended Surface ·
Plate Coil · Long Pipe Coil · Headered Coil · Accumulator or Surge
Drum · Direct-expansion Evaporator · Bottom-fed Coils · Static Head
Effect on Operation · Heat-transfer Rate · Baudelot Cooler · Protect Evap-
orator from Freezing

Chapter 7. REFRIGERANT CONTROLS   72

Where Used · Expansion Valve · Capillary Tube · Advantages of Hand
Expansion Valve · Automatic Expansion Valve · Thermostatic Expansion
Valve · Equalizers · Influence of External Equalizer · Connecting Ex-
ternal Equalizer into Circuit · Multi-outlet Thermostatic Expansion Valve
· Remote Bulb · Power Assembly · Remote-bulb Charges · Selecting
Thermostatic Expansion Valve · Remote-bulb Location · Float Valves
· Low-side Float Construction · Advantages of Low-side Float Valves ·
Problems with Low-side Floats · High-side Float Valve · Purge Valve
Need · Float Switch · Flooded Evaporator Control · Preventing Liquid
Seal Loss · Precautions with Float Switches · Float Switch Piping · Sole-
noid Valves · Solenoid Valve Use · Solenoid Valve Installation Precau-
tions · Suction Line Regulators · Evaporator Pressure Regulator · Pilot-
operated Evaporator Pressure Regulator · Evaporator Pressure Regulator to
Control Temperatures · Air-modulated Evaporator Pressure Regulator ·
Suction Hold-back Valve · Automatic Valve Precautions · Pilot-operated
Control for Suction Stop · Liquid-line Shutoff · Precautions When Selecting Solenoid Valve Power Supply · Strainers · Manual-opening Stem on
Solenoid Valve · How to Test Controls

Chapter 8. ELECTRIC CONTROLS   96

Three Types of Pressure Controllers · Temperature versus Pressure Con-

trollers · Bimetal Thermostat · Range of Controllers · Adjustable Differ-
ential Thermostat · Remote-bulb Thermostat · SPST, SPDT, DPST,
DPDT · Low Oil-pressure Switch · High-pressure Cutouts · Reset Pressure
Controllers · Floating Control · Differential Controller · Float Switches
· Installing Electric Controls · Troubleshooting

Chapter 9. PIPING AND FITTINGS.   107

Velocity and Pressure Drop · Pressure Drop from Receiver to Expansion
Valve · Pressure Drop in Suction Line · Piping for Freon-12 Plants ·
Piping Joints for Freon · Connections on Liquid Receiver · Hot-gas Line
· Suction Piping Hookups to Compressor · Expansion Valve Hookup ·
Piping for Room Cooler · Piping for Parallel Operation · Backpressure
Regulator Hookup · Thermostatic Expansion Valve Hookup · Flooded-
coil Hookup · Shell-and-tube Liquid Chiller Hookup · Hookup of Several
Evaporators · Solenoid Valve Hookup for Brine System · Solenoid Valve
for Compressor Capacity Control · Table of Materials for Refrigeration
Piping · Valves · Insulation for Low Temperatures · Properties to Con-
sider · Thermal Conductivity · Vapor Barrier to Prevent Moisture Harm-
ing Insulation · Reflective Surfaces

Chapter 10. LUBRICATION   124

Lubrication of Moving Parts · Force-feed Lubrication · Splash-type Sys-
tem · Controlling Splash Lubrication · Oil in Evaporator · Advantages of
Force-feed Lubrication · Lube System of Freon or Ammonia · Cylinder
Lubrication · Double-trunk Piston · Starting a Compressor · Overexpansion
· Foaming · Oiling Bearings · Cylinder Walls · Oxidation · Oil Specifi-
cations · Classifying Petroleum Lubricants · Reducing Friction · Lube
Tests · Viscosity · Pour Point · Flash and Fire Point · Cloud Point ·
Halide Refrigerants · Lube Oil for Ammonia · Sulfur Dioxide and Lube ·
Methyl Chloride and Lube · Freon and Lube · Sludge · Deposits ·
Excessive Consumption · Care · Adding Oil to Crankcase · Centrifugal
Compressor Lubrication · Storage and Handling · Dispensing · Jam-type
Packing · Mechanical Seal · Free-floating Packing · Bellows Seal · Lan-
tern Rings

Chapter 11. DEFROSTING   141

Frost Effect on Coils · Insulating Quality of Frost · Simple Defrosting ·
Precautions · Hot-gas Method · Warm Brine Lines · Semiautomatic Device · Full-automatic Device · Reverse Flow · Defrosting Ammonia Systems · Electrical Wiring Hookup · Electric Resistance Heating

Chapter 12. COMPRESSOR MOTORS AND DRIVES   146

Two Types of A-c Motors · Single-phase versus Polyphase · Squirrel-cage
Induction Motor · Starting Squirrel-cage Motors · Slip-ring Motor ·
Wound-rotor Motor · Single-phase Motors · Hermetically Sealed Motor
· Synchronous Motor · Motor Enclosures · Motor-Compressor Hookups
· Steam Drives · Internal-combustion Engine Drives · V Belts · Break-

away Torque · Accelerating Torque · Pull-up Torque · Pull-in Torque
· Pull-out Torque · Motor Controls

Chapter 13. PRACTICAL CALCULATIONS   159

Horsepower Equivalent · Ice-making Capacity · Refrigeration Rate · Refrigeration Needed to Cool Beef · Loss through Walls · Equation of Heat
Loss · Heat Transmission Problem · Heat Generated in Storage · Cooling
Water Needed · Refrigerant Effect for Freon · Heat Removed from Storage · Calcium Chloride Needed · Ice-making Capacity versus Refrigeration Capacity · Ammonia Leakage into Brine · Lost Cooling Effect

Chapter 14. OPERATION   175

Thermometers and Efficiency · "Stick Finger Test" · Weight of Refrigerant
· Discharge Pressure · Ammonia Drum Sizes · Charging Large Ammonia
System · Ammonia in System · Withdrawing Ammonia · Charging Freon
System · Reverse Flow Dangers · Starting Reciprocating Compressor · Air
in System · Keeping Load Balanced · Purge Centrifugal System · Surg-
ing · Operating Absorption System · Low Suction Pressures · Starting
Ammonia System · Shutting Down Ammonia System · Foul Gas · Auto-
matic Purger · Hand Purging · Purging Centrifugal Compressors · Ex-
pansion Valve Malfunction · Low-Temperature Operation · Testing for
Leaks · Halide Torch · Test Safety Valves · Moisture and Freon · Driers
· Removing Cylinder Head · Checking Cylinders and Rings · Wear on
Trunk Pistons · Discharge Valve Failure · Removing Moisture · Accidents
with Reciprocating Compressors · Troubleshooting

Chapter 15. REFRIGERATION APPLICATIONS   198

Cold Effect on Food · Temperature Ranges for Food · Quick Freezing ·
Sharp Freezer · Blast Freezer · Contact Freezing · Immersion Freezing ·
High Humidity · Precooling Fruits · Heat of Respiration · Controlled
Atmosphere Storage · Ice Making · Dry Ice · White Ice · Refrigeration
in Chemistry

Chapter 16. COOLING TOWERS   207

Removing Heat · Reusing Water · How Tower Works · Atmospheric
Water Cooling · Heat from Water to Air · Cooling by Evaporation · Wet-
and Dry-bulb · Air for Cooling · Speeding Cooling Effect · Properties
of Air · Cooling Range · Approach · Heat Load · Pumping Head · Drift
· Blowdown · Make-up · Cooling Pond · Spray Pond · Filling Material
· Classifying Cooling Towers: (1) Atmospheric, (2) Forced-draft, (3)
Counterflow Induced-draft, (4) Double-flow, and (5) Hyperbolic Evapora-
tive Condenser · Precautions · Spotty Performance · Prevent Ice Forma-
tion · Fog · Winter Operation · Running Fan in Reverse · Mechanical
Maintenance · Structural Maintenance · Shutting Down · Limitations ·
Sizing Pump · Water per Minute · Automatic Sprinklers · Increasing
Capacity · Protection from Freezing · Spring Startup · Maintenance ·
Laying up for Winter

Contents

Chapter 17. AIR CONDITIONING 227

Definition · Conditioning Load · Heat Transmission · Outside Temperatures · Sunlight · People Load · Infiltration · Air Saturation Temperature · Relative Humidity · Dew Point · Wet Bulb · Latent Heat · Sensible Heat · Total Heat · Psychrometric Chart · Air Mixtures · Air Volume · Spray Cooling · Jet Cooling · Absorption System · Compression Cycle · Controlling Temperature · Human Effect on Air · Removing Odors · Spray Humidifying · Removing Dust · Air Mixtures · Packaged Units · Conditioning Old Buildings · Code and Sanitary Regulations · City Codes · Ice System · Ground Water · Air Blending · Central Air-conditioning System · Zoning · High Velocity · Reverse Operation · Controlling Compressor Capacity · Computerized Control · Vibration Isolation · Avoiding Drafts · Sprayed Roof · Troubleshooting Calculations · Starting Centrifugal Compressor Systems · Shutdown of Centrifugal Systems

Chapter 18. COOLING WATER AND BRINE 263

Circulating Systems · Open and Closed Systems · Langelier Index · Control Scale Method · Surface-active Materials · Corrosion · Slime and Algae · Chemicals · Intermittent Feed · Wood Destruction · Delignification · Scale Deposits · Fungus Attack · Alkalinity · Blowdown Calculations · Brine Systems · Brine Coolers · Kinds of Brine · Density of Brine · Ammonia Leakage in Brine · Testing Brine · Meaning of pH

Chapter 19. SAFETY 272

Refrigerant Code · Explosions · Gage Glass · Safety Rule · Keeping Machinery from Turning · Precautions · Explosions · Fire Extinguishers · Gas Mask · Smoking · Periodic Testing · Toxic Refrigerants · Ammonia Dangers · Ammonia Leak · Ammonia in Eyes · Ammonia and Flames · Carbon Dioxide · Halocarbon Dangers · Carbon Tetrachloride · Gasoline Dangers · Safe Solvent · Dry Ice · Low Voltage · Dumping Ammonia · Preventing Slugs of Ammonia · Oil in System · Oxygen for Leak Test · Canister Gas Mask · Test Pressures · Checking New System

Chapter 20. CRYOGENICS 287

Cryogenics New Field · Two Major Classifications · Gases Liquefied · How Air Is Liquefied · Joule-Thomson Effect · Expansion Engine · Claude System · Cascade System · Component Gas Separation

Chapter 21. MACHINERY MANAGEMENT 292

Training · Log Sheets · Shift Schedules

Chapter 22. LICENSE REQUIREMENTS FOR REFRIGERATION AND AIR CONDITIONING ENGINEERS IN THE UNITED STATES AND CANADA .. 295

Index 313

1

FUNDAMENTALS

Q   What is refrigeration?
A   Refrigeration has many definitions: (1) the development in a given space of a temperature lower than that which exists in some other or adjacent space, (2) the process of cooling or removing heat, or (3) the process by which "cold" is produced. Actually, refrigeration is all these things, but more specifically it is (4) the process of removing heat from a space or substance to reduce its temperature and transferring that heat to another space or substance.

Q   What is heat?
A   Heat is a form of energy, a distinct and measurable property of all matter. The quantity of heat depends on the quantity of the substance and the type of substance involved. Heat cannot be destroyed but can be transferred from one substance to another, always moving from the warmer to the colder substance. On this fact the science of refrigeration is based.

Q   What is temperature?
A   See Fig. 1-1. Temperature is the measure of the *intensity* or level of heat. The unit of temperature most commonly used is the degree Fahrenheit (°F). One degree Fahrenheit is 1/180 of the difference between the boiling point of water and the melting point of ice under standard atmospheric pressure. On the Fahrenheit scale, the boiling point of water is 212°, and the melting point of ice is 32°. Other temperature scales are (1) the centigrade scale on which the melting point of ice is taken as 0° and the boiling point of water is 100° and (2) the Reaumur, on which the melting point of ice is 0° and the boiling point of water is 80°. Still other scales are used to measure the intensity of heat.

Q   What is the unit of quantity of heat?
A   The unit used to measure the quantity of heat is the British thermal unit, more commonly known by its abbreviation Btu. The Btu is the amount of heat energy required to raise the temperature of one pound of

1

water one degree Fahrenheit. Conversely, if the temperature of one pound of water is reduced one degree Fahrenheit, one Btu of heat energy is removed.

By this definition the exact value of one Btu depends upon the initial temperature of the water. Usually, the *mean Btu,* which is 1/180 of the heat required to raise the temperature of one pound of water from 32 to 212°F at a constant atmospheric pressure of 14.696 lb per sq in. absolute is used.

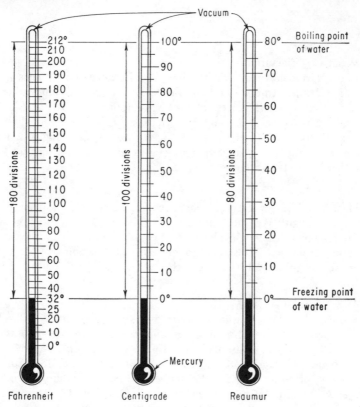

Fig. 1-1. Three of many thermometer scales used to measure the intensity of heat.

Q What is specific heat?

A The specific heat of a substance is the amount of heat needed to raise the temperature of one pound of that substance one degree Fahrenheit. Since it requires one Btu to raise the temperature of one pound of water one degree Fahrenheit, the specific heat of water is 1.0. Because most other substances need less heat energy to create a given change in temperature, they have a specific heat of less than 1.0 (see Fig. 1-2). To find the heat required to change the temperature of any substance,

Item	Specific heat, Btu per lb
Most fruits and vegetables.........	0.92
Meats:	
Beef........................	0.77
Fish, poultry..................	0.82
Lamb, pork..................	0.66
Dairy products:	
Milk........................	0.92
Butter.......................	0.64
Eggs........................	0.76

Fig. 1-2. Almost every piece of matter has a different specific heat.

multiply its weight in pounds times its specific heat times the temperature change required, or

$$H = WS \ (t_2 - t_1)$$

where H = heat, Btu
W = weight, lb
S = specific heat
t_1 = original temperature, °F
t_2 = final temperature, °F

EXAMPLE: How much heat is needed to **raise the temperature** of 200 lb of butter from 40 to 82°F?

$$H = WS \ (t_2 - t_1)$$
$$H = 200 \times 0.64 \times (82{-}40)$$
$$H = 200 \times 0.64 \times 42$$
$$H = 5376 \text{ Btu of heat required}$$

Conversely, if it is required to cool a product, the same formula applies.

EXAMPLE: How much heat must be removed from 500 lb of beef to cool it from 94 to 34°F?

$$H = WS \ (t_2 - t_1)$$
$$H = 500 \times 0.77 \times (34°F - 94°F)$$
$$H = -23{,}100 \text{ Btu of heat}$$

Since the answer is a minus number, it indicates heat to be removed.

Q What is sensible heat?

A   Sensible heat is that heat added to or removed from a substance that can be measured by a change in the temperature of the substance.

Q   What is latent heat?
A   Latent heat is that heat added to or removed from a substance to cause a change of state without a change of temperature. All substances may exist in three different conditions or states; (1) the solid state, (2) the liquid state, and (3) the gaseous or vapor state. A familiar illustration is water. At temperatures between +32 and +212°F, water is a liquid; at temperatures above 212°F, a vapor (steam); and at temperatures below +32°F, a solid (ice). Experiments have shown that 144 Btu of heat

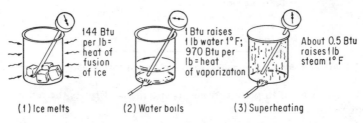

144 Btu
per lb =
heat of
fusion
of ice

1 Btu raises
1 lb water 1° F;
970 Btu per
lb = heat
of vaporization

About 0.5 Btu
raises 1 lb
steam 1° F

(1) Ice melts   (2) Water boils   (3) Superheating

Fig. 1-3. Matter exists in three states, depending on temperature.

energy must be added to 1 lb of ice to change it to 1 lb of water. During the addition of this amount of heat, the temperature remains constant at 32°F. That is, a change of state has occurred without a change of temperature. Further, if heat is continually added, 180 Btu will be needed to raise the temperature of the pound of water to 212°F. At this point, the water will begin to boil or change to steam (vapor). An added 970 Btu will be required to change the pound of water to steam. However, as long as any liquid remains, the temperature will not be raised above 212°F (see Fig. 1-3).

Q   Explain latent heat of fusion; vaporization; and condensation.
A   The latent heat of fusion is the amount of heat needed to change a substance from the solid to the liquid state. It is usually expressed in Btu per pound. The latent heat of vaporization is the amount of heat required to change a substance from the liquid to the vapor state. It is also expressed in Btu per pound. The latent heat of condensation is the amount of heat that must be removed from a vapor to change it to a liquid (condense it) and is the same as the latent heat of vaporization. For example, 970 Btu must be added to 1 lb of water to change it to 1 lb of steam. If the pound of steam is to be returned to a liquid, or condensed, 970 Btu must be removed.

Q   What is energy?

A   In the simplest terms, energy is the ability to perform work. It may exist in several forms, such as heat energy, mechanical energy, chemical energy, or electrical energy, and may be changed from one form to another. For example, chemical energy in a storage battery becomes electrical energy flowing through a circuit that lights a lamp (light energy and heat energy) or turns a motor (mechanical energy). Though it may be changed from one form to another, energy cannot be created or destroyed, so the same relationships of energy transformation always apply (see Fig. 1-4).

1 horsepower (hp)	= 33,000 foot-pounds (ft-lb) per min
	= 550 ft-lb per sec
	= 746 watts
	= 2545 British thermal units (Btu) per hour (hr)
1 horsepower-hour (hp-hr)	= 1 hp for 1 hr
	= 746 watthours (whr)
	= 0.746 kilowatthours (kwhr)
	= 2545 Btu
1 kilowatt (kw)	= 1,000 watts
	= 1.34 hp
1 kilowatt-hour (kwhr)	= 1 kw for 1 hr
	= 1,000 whr
	= 1.34 hp-hr
	= 3415 Btu
1 British thermal unit (Btu)	= 778 ft-lb

Fig. 1-4. Energy may be changed from one form to another.

Q   Why is an understanding of the laws of energy important to those working with refrigeration?

A   Refrigeration is the process of transferring heat from one area to another. Since heat is one of the common forms of energy (all other forms of energy may be entirely converted into heat energy), understanding some of the basic physical laws of energy helps in understanding refrigeration. Regardless of its form or source, all energy entering into a refrigeration system must finally be balanced by energy leaving the system.

Q   Explain how some of the more common forms of energy are converted into heat energy that affects a refrigeration system.

A   In a brine circulation system, electrical energy is converted into mechanical energy in a motor used to drive a circulating pump. All of the mechanical energy that is used to circulate the brine against pipe friction, to overcome pressure drop, etc., is converted into heat energy. In a system using motor-driven fans for circulating air, the energy used to drive the fans is all converted into heat energy. If the motors are located within the conditioned space, all the electrical input to the motor, including motor losses, is converted into heat energy. Refrigerant gas flowing through a suction line creates a small amount of heat energy in overcoming pipe friction and in causing other gas to flow in the line. This is represented by the pressure drop in the line and is proportional to the velocity of the gas. The lights in any refrigerated space represent 3.4 Btu of heat for each watt of electrical energy consumed. These are but a few examples of heat energy that enter a system. They must be added to heat removed from the air or other substances to be cooled.

Q   What is absolute zero?
A   All bodies are known to be made up of a large number of small particles known as molecules. Molecules are in constant motion, vibrating to and fro. The faster they move, the hotter the body. On the Fahrenheit scale 459.8°F below zero (−460°F) is known as absolute zero. On the centigrade scale absolute zero is 273.2°C below zero. At this point there is absolutely no vibration of molecules; there is therefore no heat.

Q   What is the kinetic theory of heat?
A   According to the kinetic theory of heat, the phenomena of heat are the result of the vibrating energy of the molecules of any substance. As heat is added to a substance, it causes the molecules to move more rapidly as the heat content and temperature rise. If enough heat is added, molecular activity increases until a solid, such as ice, becomes liquid, or water. If heat continues to be added to the liquid, temperature and molecular activity continue to rise until the boiling point of the liquid is reached. At this point, enough energy has been added to the liquid to make molecular activity so great that the liquid boils and large numbers of molecules escape from the liquid. The large amount of heat needed to create this boiling action (or the conversion of a liquid into a vapor) is the latent heat of vaporization. This is the basis for mechanical refrigeration.

Q   What effect does pressure have on the boiling point of a liquid?
A   All liquids have a given boiling point for a given pressure condition. In other words, all fluids behave like water in that, when in the liquid state, they evaporate as soon as heat is added. And like water, when in the vapor state, they condense when heat is taken away, at a temperature known as the saturation temperature. Thus the saturation temperature for any fluid depends upon the pressure on the fluid. For example, water at sea-level conditions of 14.7 pounds per square inch (psi) pressure on the surface must be heated to 212°F to boil or for the latent heat of vaporization to occur. If the pressure is increased, the boiling point is increased. If the pressure is decreased, the boiling point is decreased.

Q   How is this principle used in refrigeration?

A Refrigeration systems use liquids with boiling points that may be many degrees below 0°F and whose boiling points at different pressures, or pressure-temperature characteristics, are known. Through the use of various mechanical devices, the pressure within a system can be maintained at any desired point. Since the pressure can be controlled, the temperature may be controlled as well.

Q Define absolute pressure and gage pressure.
A Absolute pressure (psia) is the pressure in pounds per square inch above a complete vacuum. Gage pressure (psig) is the pressure in pounds per square inch above normal atmospheric air pressure of 14.696 psi.

Q What is the relationship between "inches of vacuum" and absolute pressure?
A See Fig. 1-5. Any pressure below 0 lb gage is sometimes referred to as so many inches of vacuum. Since normal atmospheric pressure of 14.696 lb will support a column of mercury 29.92 in. in height, a 10 in. vacuum refers to a vacuum or partial pressure of 10 in. less than normal pressure. A complete vacuum, or 0 lb absolute (0 psia) is referred to as a 30 in. vacuum for practical purposes.

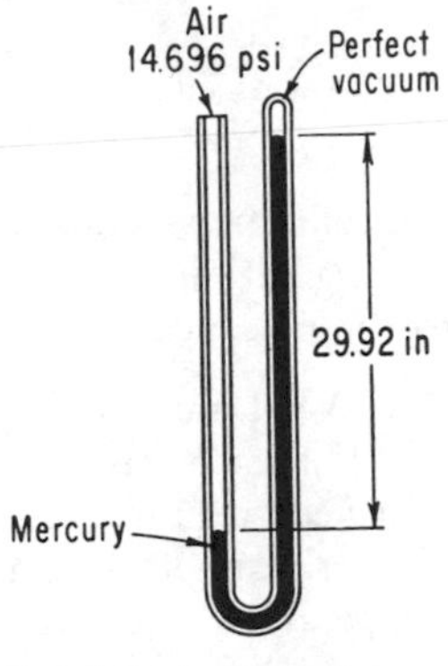

Q How is inches of vacuum converted to pounds pressure absolute?
A To change a vacuum to absolute pressure, the reading in inches of vacuum is subtracted from 30, and the result is multiplied by 0.49. For example, a 10 in. vacuum would be 30 − 10, or 20 psia × 0.49 = 9.8 psia.

Fig. 1-5. Atmospheric pressure supports a mercury column 29.92 in. high at sea level.

Q Which term is generally used in refrigeration, psia or psig?
A Most pressure gages read in pounds per square inch gage. Calculations to determine compression ratios, etc., must be based on pounds per square inch absolute. Absolute pressure may be found easily by adding 14.696 (usually taken as 14.7) lb to gage pressure.

Q What is the law of partial pressures, and how does it affect refrigeration systems?
A Dalton's law of partial pressures states that if a mixture of gases or vapors is enclosed in a container, each will exert its own pressure on the container entirely independent of the others. Then the total absolute pressure will be equal to the sum of partial pressures exerted by each of the gases. For example, if container A (Fig. 1-6) has liquid refrigerant to

a depth of 3 in. at a temperature of 86°F, the pressure will be 169.2 psia. If container *B* has enough air to register 35 psia, and valves A and B are opened while the temperature remains constant, pressure in both vessels will soon reach 169.2 + 35, or 204.2 psia. Enough liquid refrigerant will evaporate to maintain the 86°F temperature and 169.2 psia pressure. The two gases will intermix, each exerting its own pressure on

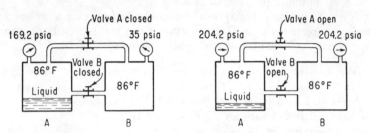

Fig. 1 6. Each of the gases or vapors in a closed container exerts its own pressure, independent of the others.

the containers. Thus the presence of air or other noncondensable gases in a refrigeration system will cause pressures greater than the pressure due to the presence of the refrigerant. For further details see Chap. 14.

Q   What are critical pressure and critical temperature?
A   For each gas there exists a temperature above which it cannot be liquefied, regardless of pressure. This is called its critical temperature. The critical pressure is the pressure that causes liquefaction at critical temperature (see Chap. 3, "Refrigerants").

Q   What is a frigorific mixture?
A   So-called frigorific mixtures are substances used in laboratory methods of producing a drop in temperature. Probably the most common example of a frigorific mixture is ice or snow and salt. Adding a foreign substance to a liquid in which it can be dissolved lowers its freezing point.

Adding 10 percent salt (NaCl), by weight, to water lowers its freezing point from 32 to 18.7°F. Thus adding this percentage of salt to snow or to finely divided ice will cause it to return to the liquid state at any temperature above 18.7°F. The ice at once begins to melt, but to do so it must absorb 144 Btu per lb. If this heat is not forthcoming, the temperature of the mixture will remain below its freezing point, and, aside from the lesser intimacy of the mixture, the combination of the two substances (ice and salt) will be identical to frozen brine of the same composition. Also, it will exist under the same temperature conditions.

Q   Explain the term 1 ton of refrigeration.

A The American Society of Refrigerating Engineers defines the standard commercial ton of refrigeration as the transfer of 200 Btu per min, or 12,000 Btu per hr. It is the basis of all refrigeration calculations, whether for cold storage, air conditioning, ice making, or ice-cream manufacture.

When a person says he has a 70-ton ice plant, that is something else. He means that he can make 70 tons of ice in 24 hr, and the rated refrigerating tonnage of his plant, as defined above, might be as high as 150.

Thus, in Btu, 1 ton of ice represents $2,000 \times 144 = 288,000$ Btu. Any apparatus that can remove this amount of heat in one day has a capacity of one ton of refrigeration. And a unit that can remove 120,000 Btu per hour has a capacity of 10 tons.

Another way to put it is that 1 ton of refrigeration is equivalent to the amount of heat it takes to melt one ton of ice in 24 hr. Thus, with the latent heat of ice equal to 144 Btu per lb, this requires 288,000 Btu in 24 hr, or 12,000 Btu per hr, or 200 Btu per min.

Q How great is the heat-absorbing capacity of a substance?

A When only its temperature is raised, and its state remains the same, a substance has a comparatively small heat-absorbing capacity. That of water in the liquid state is 1 Btu per degree rise in temperature, and it is less in both the solid and gaseous states. The specific heat of other substances in all three states is, in general, less than unity, or that of water in the liquid state.

However, the heat-absorbing capacity of a substance available in a change of state, involving its latent heat of fusion and vaporization, is comparatively large. That of water is 144 Btu for fusion, and about 966 Btu for evaporation.

At its boiling point, a substance has its greatest heat-absorbing capacity, which is called its latent heat of vaporization. Since increasing the pressure has the effect of raising the boiling point (the temperature at which the liquid vaporizes) and decreasing the pressure has the effect of lowering it, it is only natural that a reduction in pressure to below that of the atmosphere (to a partial vacuum) was first used in attempts to cause some of the better known liquids to boil at a temperature low enough to produce artificial cold.

Q What does superheat mean, and what causes ammonia to become superheated in the evaporator coils?

A Pressure and temperature of saturated ammonia have a definite relationship. The temperature corresponding to a pressure of 20 psi is 5.5°F. This should be the temperature of the gas as it leaves the evaporator

Wind speed, mph	What the thermometer reads, °F											
	50	40	30	20	10	0	−10	−20	−30	−40	−50	−60
	What it equals in its effect on exposed flesh											
Calm	50	40	30	20	10	0	−10	−20	−30	−40	−50	−60
5	48	37	27	16	6	− 5	−15	−26	− 36	− 47	− 57	− 68
10	40	28	16	4	− 9	−21	−33	−46	− 58	− 70	− 83	− 95
15	36	22	9	− 5	−18	− 36	−45	−58	− 72	− 85	− 99	−112
20	32	18	4	−10	−25	− 39	−53	− 67	− 82	− 96	−110	−121
25	30	16	0	−15	−29	−44	−59	− 74	− 88	−104	−118	−133
30	28	13	− 2	−18	−33	−48	−63	− 79	− 94	−109	−125	−140
35	27	11	− 4	−20	− 35	−49	−67	−82	− 98	−113	−129	−145
40	26	10	− 6	−21	−37	−53	−69	− 85	−100	−116	−132	−148

Little danger if properly clothed	Danger of freezing exposed flesh	Great danger of freezing exposed flesh

Fig. 1-7. Wind chill index spells out temperature effect on exposed flesh.

coils. But a thermometer in the discharge line may show a temperature of 8 or 10°F. The difference is superheat.

If we could remove the gas as soon as it formed, giving it no chance to remain in the coils and pick up heat, we would not have superheat at this point. A poorly insulated suction line also helps superheat the gas, but it is impossible to eliminate superheat entirely.

TERMINOLOGY USED IN REFRIGERATION

ABSOLUTE HUMIDITY: The weight of water vapor in a unit volume, usually expressed as grains per cu ft (7,000 grains equals 1 lb).

AIR BOUND: Air trapped in piping, equipment, etc. such as a steam radiator, which prevents maximum heat transfer; or air trapped in the suction side of a pump which causes loss of suction.

AMBIENT TEMPERATURE: The temperature of air in a space, e.g., room temperature.

ANHYDROUS: Free of water, especially water of crystallization.

BACKPRESSURE: Another term for suction pressure.

BRINE: In refrigeration systems, any liquid that is cooled by the refrigerant and pumped through the cooling coils to pick up heat. It does not undergo any change in state, but only in temperature. Brine is used in indirect systems; refrigerant is used in direct systems.

CALORIE: The quantity of heat required to raise the temperature of one gram of water one degree centigrade.

CHILLED WATER: A cooling medium that removes heat from the area to be cooled and gives up the heat in the chiller.

CHILLER: A heat exchanger in which low-pressure refrigerant boils or vaporizes, thus absorbing the heat that was removed from the refrigerated area by the cooling medium (water).

CHILLER LOAD: An indication of the number of tons of refrigerant being produced.

COEFFICIENT OF PERFORMANCE: The ratio of refrigerating effect to work of compression. A high coefficient of performance means high efficiency. The theoretical coefficients range from about 2.5 to more than 5.

COOLING MEDIUM: A fluid used for picking up heat which is circulated to the heat exchanger, where heat is removed; examples are chilled water and brine.

DEGREE-DAY: For any given day, the number of heating degree-days is the difference, in degrees, between the average (mean) temperature for that day and 65°F. For example, if the mean temperature for a day is 40°F, then the number of degree-days for that day equals 65 − 40 = 25 degree-days. Thus, when the mean temperature is less than 65°F, there are as many degree-days as there are Fahrenheit degrees difference between the mean temperature for the day and 65°F.

DEHUMIDIFY: To reduce the quantity of water vapor within a space.

DEHYDRATE: To remove water from any form of matter.

DEW POINT: The temperature at which the water vapor in the air begins to condense, or the temperature at which the relative humidity of air becomes 100 per cent.

ENTHALPY: The total heat or heat content of a substance, expressed in Btu per lb.

FLOODED REFRIGERATION SYSTEM: A type of system where only part of the circulated refrigerant is evaporated, with the remainder being separated from the vapor and then recirculated.

FREEZE-UP: Ice formation on a refrigeration system at the expansion device, making the device inoperative.

HEAD PRESSURE: Pressure at the discharge of a compressor or in the condenser. It is also known as "high-side" pressure.

HIGH SIDE: The portion of a refrigeration system that is under discharge or condenser pressure. It extends from the compressor discharge to the expansion valve(s) inlet.

HORSEPOWER PER TON: Mechanical input in horsepower, divided by tons of refrigerating effect produced. If the coefficient of performance is known, the horsepower per ton can be figured directly; divide 12,000 Btu per hr by 2545 Btu per hp-hr and the coefficient of performance.

LATENT HEAT OF FUSION: The heat added or extracted when a substance changes from the solid to the liquid state or from the liquid to the solid state. For example, when ice melts in a refrigerator, 144 Btu per lb must be added to produce the melting, or when ice freezes in an ice tank, 144 Btu per lb must be extracted; 144 Btu per lb is the latent heat of fusion.

LIQUID LINE: Refrigerant piping through which liquid refrigerant flows from the condenser to the expansion valves.

LOW SIDE: The portion of a refrigeration system in which the refrigerant is at low pressure. It extends from the expansion valve(s) outlet to the suction inlet of the compressor.

MECHANICAL EQUIVALENT OF HEAT: One Btu equals 778.2 ft-lb of mechanical energy.

PLENUM CHAMBER: An air compartment maintained under pressure with connections to one or more distributing ducts.

PUMP-DOWN: The operation by which the refrigerant in a charge system is pumped in liquid form into the condenser/receiver.

REFRIGERANT HANDLED: The amount of refrigerant circulated. Dividing 200 Btu per min by the refrigerating effect, in Btu per lb of refrigerant, gives the number of pounds of refrigerant circulated each minute.

REFRIGERATING EFFECT: The amount of heat absorbed in the evaporator, which is the same as the amount of heat removed from the space to be cooled. It is measured by subtracting the heat content of 1 lb of refrigerant as it enters the expansion valve from the heat content of the same pound of refrigerant as it enters the compressor.

STANDARD-TON CONDITIONS: An evaporating temperature of 5°F, a condensing temperature of 86°F, liquid before the expansion valve at 77°F, and a suction gas temperature of 14°F produce standard-ton conditions. Refrigerating machines are often rated under standard-ton conditions.

SUCTION PRESSURE: Pressure in the compressor suction or at the outlet of the evaporator. It is also know as "low-side" pressure.

VOLATILE: Easily evaporated. This is a necessary property of all compression refrigerants.

WIND CHILL FACTOR: Temperature effect on exposed flesh at certain wind speeds and temperatures, as shown in Fig. 1-7. If the weather temperature is 10° above zero, for example, and the wind is blowing at 20 mph, the wind chill factor is 25° below zero.

WORK OF COMPRESSION: The amount of heat added to the refrigerant in the compressor. It is measured by subtracting the heat content of 1 lb of refrigerant at the compressor suction from the heat content of the same pound of refrigerant at the compressor discharge. Multiplying the work of compression, in Btu per pound, by the number of pounds of refrigerant handled in an hour, and dividing by 2545 Btu per hp-hr, gives the theoretical power requirements.

SUGGESTED READING

Reprints sold by *Power* magazine, costing $1.20 or less
   Power Handbook, 64 pages
   Refrigeration, 20 pages

Books
   Elonka, Steve, and Joseph F. Robinson: *Standard Plant Operator's Questions and Answers*, vols. I and II, McGraw-Hill Book Company, New York, 1959.
   Elonka, Steve: *Plant Operators' Manual*, rev. ed., McGraw-Hill Book Company, New York, 1965.
   Elonka, Steve, and Alonzo R. Parsons: *Standard Instrumentation Questions and Answers*, vols. I and II, McGraw-Hill Book Company, New York, 1962.
   Elonka, Steve, and Anthony L. Kohan: *Standard Boiler Operators' Questions and Answers*, McGraw-Hill Book Company, New York, 1969.

2

REFRIGERATION SYSTEMS

Q What were some early means of refrigeration?
A Before the advent of mechanical refrigeration water was kept cool by storing it in semiporous earthenware jugs so that the water could seep through and evaporate. The evaporation carried away heat and cooled the water. This system was used by the Egyptians and by Indians in the Southwest. Natural ice from lakes and rivers was often cut during the winter and stored in caves, straw-lined pits, and later in sawdust-insulated buildings to be used as required. The Romans carried pack trains of snow from the Alps to Rome for cooling the Emperors' drinks. Though these methods of cooling all make use of natural phenomena, they were used to maintain a lower temperature in a space or product and may properly be called refrigeration.

Q How does the use of ice in a refrigerator cool the products stored in the same space?
A Heat always flows from a warm to a cold substance or from the products stored in a refrigerator to the ice. In absorbing this heat energy, the ice is melted or turned to water. The water flows by gravity out of the refrigerator carrying the latent heat of fusion with it. Each pound of ice that melts absorbs 144 Btu of heat.

Q Name five means of producing refrigeration.
A (1) Use of natural or manufactured ice; (2) compression systems using one or more of the several refrigerants; (3) absorption systems using one or more of the several refrigerants; (4) vacuum systems using one of several means of reducing the pressure to near absolute zero; and (5) thermoelectric process using a d-c flow through thermocouples (still in the development stage).

Q Can ice be used to produce temperatures below the freezing point of water (32°F)?
A Yes. The addition of salt lowers the freezing point of water or the melting point of ice to nearly −6°F and has been used extensively for

14

shipping and storage of frozen foods. This requires large amounts of salt (up to 30 per cent by weight), and though the theoretical temperature available is −6°F, it is seldom reached in actual practice. Therefore, some mechanical means must be used for producing low temperatures.

Q What principle makes mechanical refrigeration possible?

A Mechanical refrigeration is possible because a volatile liquid, called a refrigerant, will boil under the proper conditions and in so doing will absorb heat from surrounding objects. The actual refrigerating or cooling effect is produced by the boiling refrigerant changing to a vapor, not by the machine.

Q Explain how a refrigerant produces a cooling effect.

A Figure 2-1 shows a flask of ammonia (boiling point −28°F at atmospheric pressure) inside an insulated box with a valved vent pipe leading to the atmosphere. At a temperature of 86°F, the pressure inside the flask would be 154.5 psig, and no action would take place. Opening the vent valve will cause ammonia vapor to escape, reducing the pressure. As the pressure is reduced, the boiling point is also reduced, and the liquid will continue to boil as long as enough heat is present in the liquid or in the contents of the insulated box. If the vent valve is left open until the pressure is reduced to atmospheric pressure,

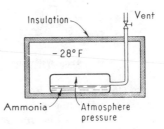

Fig. 2-1. Cooling effect is produced when ammonia boils at low temperature.

the temperature of the ammonia will be reduced to −28°F. Then the interior of the box will nearly reach this point, at which time the boiling will stop. Addition of any heat will again cause boiling of the ammonia and removal of heat.

Q What is the purpose of a compressor and other components of a refrigeration system?

A The compressor and its drive, the condenser, the receiver, the piping, and other components are merely reclaiming and control devices. They allow the refrigerant to be collected for reuse and to be controlled according to the needs of the system.

Q Draw a schematic diagram of a simple compression system, and explain the pressure cycle.

A See Fig. 2-2. The components required for a compression refrigeration system are a compressor, condenser, liquid receiver, expansion valve, and evaporator or cooling coils. The refrigerant flows from the expansion valve through the evaporator coils where it absorbs heat and becomes a gas or vapor. Then it flows to the compressor where it is compressed to the condenser pressure. In the condenser the heat is removed, and the refrigerant vapor becomes liquid and flows to the receiver. From the receiver it flows to the expansion valve to start the circuit again. The

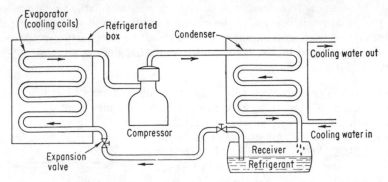

Fig. 2-2. Compression system of a refrigeration pressure cycle.

refrigerant is under low pressure from the expansion valve through the evaporator to the compressor suction. This portion of the system is called the "low side." The refrigerant is under high pressure from the compressor through the condenser, liquid receiver, and liquid line to the expansion valve. This portion of the system is called the "high side."

Q Are low sides and high sides the same on all refrigerating systems?
A No. Pressures on both sides of the systems will vary according to (1) the refrigerant used, (2) the temperature required at the evaporator, and (3) the temperature of the condensing medium.

Q Describe what temperatures might be found in the various parts of a compression system.
A See Fig. 2-3. An ammonia refrigeration system operating under standard rating conditions of 5°F evaporator temperature and 86°F condensing temperature would have the following temperatures at these points: (1) liquid refrigerant immediately before expansion valve, 77°F (9°F subcooling of liquid); (2) liquid and gas mixture immediately after passing through expansion valve, 5°F; (3) gas temperature leaving evaporator coils, 5°F; (4) suction gas temperature entering compressor, 14°F (9°F superheat gained in suction line); (5) discharge gas temperature leaving compressor, 240 to 250°F (superheated by work of compression); (6) gas temperature entering condenser, 200 to 230°F (depending on length of discharge line); (7) gas temperature in con-

denser, 86°F; (8) liquid temperature leaving condenser, 77°F (subcooled 9°F); (9) liquid temperature in receiver, 77°F. Under actual operation

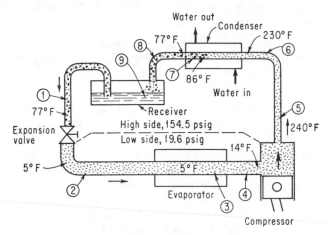

Fig. 2-3. Temperatures produced in a refrigeration pressure cycle.

there will be slight variations, but any wide variation warns an alert operator to check his system to determine the cause.

Q   What pressures would the above system have?
A   With the exception of pressure drops and piping losses, only two pressures will be found in the above system; 19.6 psig or 34.27 psia from the expansion valve, through the ammonia coils and suction line, to the compressor suction. From the compressor discharge through the condenser, receiver, and liquid line to the expansion valve, 154.5 psig or 169.2 psia.

Q   What is the compression ratio (CR) under the conditions given above?
A   The compression ratio is 169.2 psia (discharge pressure) divided by 34.27 psia (suction pressure), or CR = 4.937.

Q   What is the absorption system of refrigeration?
A   See Fig. 2-4. An absorption system makes use of the ability of one substance (the absorbent) to absorb relatively large volumes of the vapor of another substance, usually a liquid (the refrigerant). The absorbent has the ability to absorb large quantities of vapor when cold and give them up when heated.

Q   What is the most common type of absorption system in use in industrial application?
A   One of the first, and still the most widely used absorption refrigeration system today, is the system using water as the absorbent and ammonia as the refrigerant.

Q   Are there other types of absorption systems?
A   Yes, the water vapor absorption system, in which water serves as the refrigerant and a solution of lithium bromide salts in water serves as the absorbent.

Q   How does an absorption system compare with a compressor system?
A   In an absorption system (Fig. 2-4), the increase in pressure is produced by heat supplied by circulating steam or some other suitable hot gas or fluid through a coil or pipe. The absorber-generator does the work of a compressor in that the absorber replaces the suction stroke and the generator the compression stroke. The evaporator spray header corresponds to the expansion valve. The evaporator and the condenser are identical in both compression and absorption systems. The absorption refrigeration cycle utilizes two phenomena: (1) an absorption solution (absorbent plus refrigerant) can absorb refrigerant vapor, and (2) a refrigerant boils (flash-cools itself) when subjected to a lower pressure.

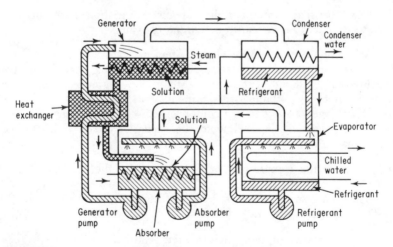

Fig. 2-4. Absorption cycle using lithium bromide solution in absorber is based on characteristics of liquid sorbents and uses heat for driving force.

Q   How do the parts of an ammonia absorption system compare with those of a compression system?

A   The essential parts of a compression system are the compressor, condenser, expansion valve, and evaporator coils (Fig. 2-3). The essential parts of an absorption system are the condenser, expansion valve, evaporator coils, absorber, and generator. In either system the condenser, expansion valve, and evaporator coils can be exactly the same. Instead of a compressor, the absorption system uses an absorber and a generator, as well as a pump to circulate the aqua (water) ammonia through the absorber and generator. In the absorber the ammonia vapor coming from the evaporator under low pressure is absorbed in the weak and comparatively cool solution of ammonia and water. In the generator heat is applied to the strong solution of ammonia and water, driving off part of the ammonia to the condenser. The weak solution is then cooled as it returns to the absorber to absorb more ammonia. This brief description shows that absorption systems and compression systems are similar in principle. The absorber takes the place of the suction stroke of the compressor by drawing the low-pressure gas from the evaporator. The generator takes the place of the compression stroke, discharging ammonia gas at high pressure and temperature. This high-pressure ammonia then passes to the condenser, is liquefied, and flows through the expansion valve of the evaporator coils just as in a compression system.

Q   What is the efficiency of the above simple absorption system, and how may it be improved?
A   The above system will perform about 300 Btu of refrigerating effect for 1630 Btu of heat input to the generator. Thus the performance factor is about 18 per cent. This may be improved by adding these auxiliary devices: (1) analyzers or bubble columns, (2) rectifiers or reflux arrangements, (3) heat exchangers, and (4) liquid precoolers. Proper application of these auxiliaries may raise the efficiency or performance factor to 70 per cent or greater.

Q   What is an analyzer or bubble column, and how does it increase efficiency of an absorption system?
A   See Fig. 2-5. The analyzer or bubble column is a pressure vessel mounted above the generator through which the vapors leaving the generator pass. Column contains a number of baffles or plates arranged with weirs that maintain a predetermined level of strong aqua fed near the top of the column. As the strong aqua overflows the weirs and flows to the next lower plate, it is heated by the hot vapors rising from the generator. As the aqua temperature increases, the ammonia vaporizes and rises through the baffles in the column. Because of the wide difference in the boiling points of water and ammonia, water present in the vapor is condensed. Water returns to the next lower plate to overflow back to the generator, while the ammonia vapor continues to the column's top.

Q   What is the rectifier, and how does it compare with a reflux?
A   See Fig. 2-6. A rectifier is the inlet part of the condenser, cooled by

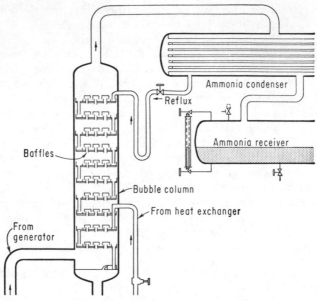

Fig. 2-5. Analyzer or bubble column heats strong aqua as it flows over the baffles or weirs.

a separate circuit of cooling water or strong aqua. It condenses a small part of the vapor leaving the analyzer and returns it as a liquid to the top baffle plate. This ensures that the vapor going to the condenser is lowered in temperature and enriched in ammonia. A reflux does about the same thing by providing a small flow of condensed ammonia liquid to the top plate of the bubble column (Fig. 2-5), but does not require a separate cooling circuit. Either system passes only commercial anhydrous ammonia vapor to the condenser.

Q   Where may heat exchangers be installed in an absorption system?
A   See Fig. 2-6. It is general practice to install a counterflow heat exchanger for the cool, strong liquid leaving the absorber and the hot, weak liquid leaving the generator. The strong liquid must be heated before it will give up the ammonia, and the weak liquid must be cooled to efficiently absorb more ammonia in the absorber. A heat exchanger at this point reduces heat input to the generator and cooling water requirements in the absorber. Liquid ammonia from the receiver may be cooled by a heat exchanger in the suction line from the evaporator. Cooler liquid

entering the expansion valve gives greater refrigerating effect per pound and increases over-all efficiency.

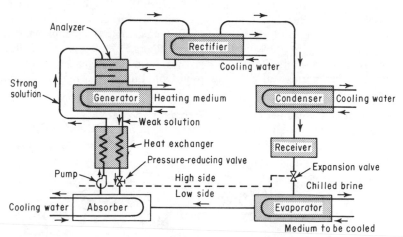

Fig. 2-6. Heat exchangers in absorption system increase the efficiency.

Q Does the ammonia-water absorption system have a low side and high side?

A Yes. From the expansion valve through the evaporator, suction line, and absorber (Fig. 2-6) the system operates under pressures similar to the low side of a compression system. That is, it provides whatever pressure is needed for the temperature. From the pump through the heat exchanger, generator, bubble column, condenser, and receiver the system operates under pressures controlled by the available condensing medium. The weak liquid returning from the generator to the absorber is controlled by a pressure-reducing valve.

Q What is a lithium bromide absorption system?

A A lithium bromide system is a water-chilling system operating on the same physical principles as an ammonia absorption system. In the lithium bromide system lithium bromide salt, which has great capacity for absorbing water vapor, is used as the absorbent, and pure water is used as the refrigerant. The basic components required are the same as for the ammonia absorption system; generator, condenser, evaporator, and absorber. The system makes use of the facts that, under a low pressure (high vacuum), water boils at a low temperature and that lithium bromide solution easily absorbs this water vapor.

Q What advantages are claimed for absorption systems over compression systems?

A   The advantages claimed are (1) savings in operating cost by using low-pressure low-cost steam—in many cases waste steam from other processes; (2) elimination of heavy electrical loads; (3) simplicity of operation and control systems that often require little attention; (4) automatic starting and stopping; (5) full efficiency at all ranges of reduced load; (6) possible installation outdoors; (7) economy of floor space required for large tonnage; (8) minimum maintenance because of fewer moving parts; and (9) minimum amount of moving equipment needed.

Q   Describe a steam-jet or vacuum refrigeration system.

A   See Fig. 2-7. Steam-jet or vacuum refrigeration is based on the principle that the boiling point of water is lowered as the pressure is reduced.

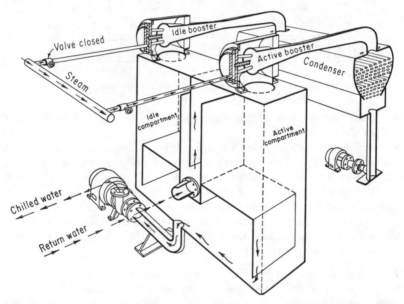

Fig. 2-7. Steam-jet system has water as the refrigerant. Ejector establishes vacuum, causing part of water to vaporize; this chills the rest.

If pressure is reduced to nearly 0 psia, the boiling point approaches 32°F. This allows for fast precooling, since the heat is actually removed by removal of a small amount of water under the high vacuum. These systems use both steam jets and vacuum pumps. A steam-jet system is also a compression system. Here water is the refrigerant, and a steam ejector instead of a compressor is used to lower the pressure and to reduce the boiling point of water. Water to be chilled is sprayed into the vessel through a series of nozzles. The steam-jet ejector draws air and vapor out of the vessel, compresses it in a converging-diverging nozzle, and sends it to the condenser at a high velocity. A vacuum of about 29.67 in. is thus formed in the vessel so that the pressure inside the vessel is only 0.13 psia. If water enters the vessel at 80°F, its boiling point at 0.13 psia becomes only 42°F (not 212°F) because the heat needed to cause this evaporation is taken from the water in the vessel. Thus, its temperature is chilled to 42°F under these conditions. Secondary ejectors remove the last traces of air from the evaporator. The secondary condenser has two separate sections, upper and lower, with a jet ejector for each part. Water can be cooled to the freezing point with this system instead of to 42°F as described here, but the cost of the high-pressure steam needed is not economical. Water as a refrigerant is very safe and widely used in air-conditioning systems.

Q   Explain the booster cycle. Why is it used?

A   See Fig. 2-8. A booster cycle is an efficient way to reduce temperature down to subzero, which is needed for quick-freezing foods, etc. One

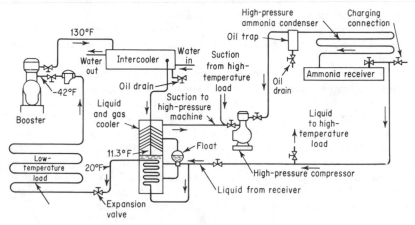

Fig. 2-8. Booster system produces low temperatures because it can handle larger volume.

pound of ammonia at 0 psig has a volume of 18 cu ft. But, if the same pound is pumped down to 10 in. Hg, its volume increases to 26 cu ft. At 15 in. Hg the volume almost doubles to 36 cu ft. As pressure corresponds to temperature, standard compressors can't efficiently handle loads from about −28 to −100°F because the volume of the gas is too large. Here is how the booster system solves the problem. Low-temperature suction gas at 10-in. vacuum, corresponding to the standard temperature of −42°F, is drawn into the suction of the booster compressor. It discharges at 25 psi at 130°F to a water-cooled intercooler. Circulating water in the tubes of this shell removes some of the heat of compression, cooling the discharge gas to a temperature dependent on the temperature and quantity of water. Gas discharges into the liquid-and-gas precooler, bubbling through the liquid that is held at a constant height in the precooler by a float control. This bubbling process cools the gas to saturation temperature corresponding to 25 psi at 11.3°F. The liquid-and-gas precooler also cools liquid ammonia coming from the receiver (which is warm) to perhaps 20°F before passing it to the low-pressure evaporator. The high-pressure compressor handles (1) gas discharged from the booster compressor, (2) gas that cools the liquid ammonia, (3) gas that cools the discharge gas from the booster, and (4) high-temperature loads, such as ice making.

Q Can a booster system be used with almost any refrigerant?
A Yes, but results show that ammonia is most suitable for holding temperatures as low as −75°F. When −100 to −110° is needed, Freon-22 is better because it works with a higher suction pressure. With Freon three stages are used for temperatures below −80°F.

Other refrigerants, like ethylene, are needed for very low temperatures. These are often condensed by Freon, or by ammonia in a cascade system.

Boosters usually pay off when gas is compressed to more than 9½ times its original pressure. When figuring the compression ratio, one must be careful to use absolute pressures, not gage pressures. As a rule, a plant carrying a suction pressure of less than 5 psig saves money with a booster. Also, a booster system takes fewer horsepower per ton of refrigeration than a single-stage compressor whenever the temperature is very low or the condensing water warm. At 0-psig suction, a single-stage unit takes 2.8 hp per ton; with a booster, only 2.2 hp per ton is used. Boosters also prevent dangerously high discharge temperatures resulting from superheat, and lubrication of the compressor is easier.

Q Can any refrigerant be used with a booster system?
A Yes.

Q How can you tell if a plant needs a booster system?
A A plant needs a booster system if the compression ratio is greater than 9.5. Use absolute rather then gage pressures when figuring compression ratio. A plant using suction pressure of less than 5 psig usually saves money with a booster system.

Q What is meant by compounding?
A See Fig. 2-8. Compounding is the same operation as that performed by a booster system.

Q   How are extremely low temperatures obtained?
A   See Fig. 2-9. Extremely low temperatures can be obtained in several

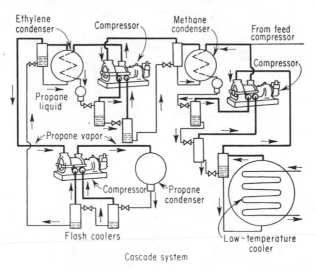

Cascade system

Fig. 2-9. Cascade system uses several refrigerants, attains very low temperatures

ways. One method is the cascade system. Here, a series of refrigerants having progressively lower boiling points are used. One refrigerant serves as the coolant to condense the refrigerant gas with the next lower boiling temperature. Cascading is usually used to obtain temperatures of below −135°F.

Q   Describe the centrifugal refrigeration unit shown in Fig. 2-10.
A   The centrifugal refrigeration unit shown consists basically of a centrifugal compressor, a cooler, and a condenser. The compressor uses centrifugal force to raise the pressure of a continuous flow of refrigerant gas from the evaporator pressure to the condenser pressure. The compressor handles large volumes of gas, and thus can use refrigerants having high specific volumes. The cooler is usually in the shell side. The condenser is usually a shell-and-tube type which uses water as a means of condensing; it may be an air-cooled or evaporative condenser for special applications.

The compressor in Fig. 2-10 is of the hermetic type, in which gas flows through the electric motor windings to the suction side of the compressor impellers. These machines may be driven at motor speed or, by means of a speed-increasing gear between the motor and compressor, at a single higher speed. Most machines have inlet guide vanes for capacity control.

Hermetic packaged units of this type have up to 2,000 tons capacity. The open type (driver outside casing) is driven at speeds of 3,000 to 18,000 rpm by an electric motor, steam turbine, gas engine, gas turbine, or diesel engine. Capacity can be varied to match the load by means of a constant speed drive with variable inlet guide vanes or suction damper control, or a variable speed drive with suction damper control.

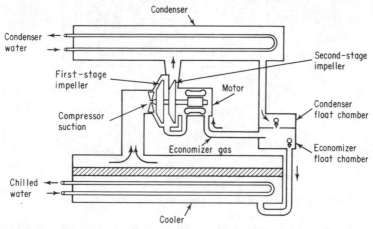

Fig. 2-10. Centrifugal refrigeration system is hermetic type with capacity up to 2,000 tons in single unit.

SUGGESTED READING

Reprints sold by *Power* magazine, costing $1.20 or less
  Refrigeration, 20 pages
  Conditioned Air for Industry, 16 pages

Books
  Elonka, Steve, and Joseph F. Robinson: *Standard Plant Operator's Questions and Answers*, vols. I and II, McGraw-Hill Book Company, New York, 1959.
  Elonka, Steve: *Plant Operators' Manual*, rev. ed., McGraw-Hill Book Company, New York, 1965.
  Elonka, Steve, and Alonzo R. Parsons: *Standard Instrumentation Questions and Answers*, vols. I and II, McGraw-Hill Book Company, New York, 1962.

3

REFRIGERANTS

Q   What is a refrigerant?
A   A refrigerant is the substance used for heat transfer in a refrigerating system. It picks up heat by evaporating at a low temperature and pressure and gives up this heat by condensing at a higher temperature and pressure.

Q   What are some of the characteristics desired in a refrigerant?
A   Many points should be considered; these are the most important: (1) Since refrigeration takes place by the evaporation of a liquid, a refrigerant must be volatile, or capable of being evaporated. (2) The latent heat of vaporization should be high enough so that circulation of a minimum amount of refrigerant will accomplish the desired result. (3) Safety in use under operating conditions is important. Refrigerants should not burn or support combustion or be explosive. (4) The refrigerant should be harmless to people and should have an odor that will make its presence known. Leaks should be detectable by simple tests. (5) Cost of the refrigerant should be reasonable, and it should be available in sufficient quantities for commercial use. (6) The refrigerant should be stable, with no tendency to break down or decompose under operating conditions. (7) It should have no harmful effect on metals or lubricants used in compressors and other components. (8) The refrigerant should have reasonable evaporating and condensing pressures. (9) It should produce as much refrigeration as possible for a given volume of vapor handled by the compressor. (10) A minimum of power should be needed for compression to the condensing temperature. (11) Critical temperature should be well above condensing temperature.

Q   In the technical sense, can water be used as a refrigerant?
A   Yes, water can be used as a refrigerant in some types of refrigeration systems. These are usually systems where the temperatures required are above 35°F. In order for water to evaporate at these low temperatures, a high vacuum must be maintained. Early systems used steam jets for this purpose. Newer systems are based on the absorption principle and

are finding increasing use in air-conditioning work (see discussion on lithium bromide systems, Chap. 2).

Q   Why is a high latent heat of vaporization desirable in a refrigerant, and what effect does it have on a refrigeration system?
A   Most of the heat handled by a refrigeration system is involved with latent heat. A high latent heat of vaporization means that a small amount of refrigerant will absorb a large amount of heat. This in turn requires less refrigerant circulating in a system to provide a given tonnage.

Q   What is meant by the density of the refrigerant vapor, and how does it influence the capacity of a system?
A   The density of a refrigerant vapor is given in the tables as the cubic feet of vapor per pound. Since the capacity of a system is a function of the pounds of refrigerant circulated per minute, it is usually best to have a refrigerant whose vapor density is high (the number of cubic feet per pound is low). This allows use of smaller suction lines and compressors to handle a given weight of refrigerant. For example, to maintain an evaporator temperature of 40°F with water as a refrigerant, a compressor would have to pump 477 cu ft of vapor per minute per ton of refrigeration. To maintain the same temperature with ammonia, a compressor would have to handle 3.4 cu ft per min (cfm) per ton.

Q   What is the drawback of a completely odorless refrigerant?
A   If the refrigerant has no odor, large amounts can escape through leaks in the system and not be noticed until the system fails to operate properly.

Q   What is the critical temperature of a refrigerant?
A   Any gas or vapor has a maximum temperature at which it may be condensed into a liquid. Above this temperature it remains a gas regardless of the pressure applied.

Q   Give an example of critical temperature and its effect on a refrigeration system.
A   The critical temperature of carbon dioxide (CO_2) is 87.8°F. If condensing water is at or above this temperature, it is impossible to condense the carbon dioxide gas. Therefore it is not practical to operate a CO_2 system where reasonably cool condensing water is not available.

Q   What are the advantages of having a refrigerant that boils at or above atmospheric pressure?
A   If refrigerant vapor must be pumped by an open-type compressor (not hermetically sealed), the crankshaft must extend through the compressor body. If the temperature desired in the evaporator requires a vacuum, air and atmospheric moisture can leak into the system through

the shaft seal. In systems using hermetic compressors, any leak in the system would allow the same type of leakage. The air will circulate through the system as a noncondensable gas, increasing operating costs. Moisture in combination with most refrigerants and oils has harmful effects on system components.

Q What is meant by stability of refrigerant?

A Stability means that the refrigerant remains in its original chemical form under the conditions imposed by the operation. Refrigerants are chemical compounds of two or more elements; for example, ammonia is a chemical combination of nitrogen and hydrogen (NH_3), which makes a good refrigerant. If the ammonia separates into a mixture of nitrogen and hydrogen gases, refrigerating effect is destroyed. The same applies to any refrigerant.

Q What are the physical properties of a refrigerant?

A Figure 3-1 lists the physical properties of the commonly used refrigerants. The physical properties of any substance are the known properties of that substance, such as weight, density, boiling point, freezing point, or condensing pressure, usually given for a stated pressure and temperature.

Q What information of value may be learned from a table of physical properties?

A Tables of this type are very useful to both the design engineer and the operating engineer. Boiling and freezing points and critical temperatures set limits of operation. Pressure characteristics determine the type of equipment and piping. Volume characteristics sometimes determine the types of compressors that must be used.

Q What is a vacuum refrigerant?

A Any refrigerant that exists as a liquid under normal atmospheric pressure and temperature must be vaporized in an evaporator under a pressure below atmospheric, or a vacuum. These are sometimes referred to as vacuum refrigerants.

Q What is a halogenated hydrocarbon?

A A halogenated hydrocarbon (or halocarbon) is any one of a group of new refrigerants that have been developed since about 1925 to overcome the irritating or toxic effects of refrigerants, such as ammonia and sulfur dioxide and the high condensing pressures required with carbon dioxide.

Q What are "Freon" refrigerants?

A "Freon" is the trade name of refrigerants manufactured by the Freon Products Division of E. I. du Pont de Nemours & Company. Because they

Name of refrigerant	Operating data													
	Chemical symbol[a]	Evaporator temperature range[b]	Boiling point, °F (atmospheric)	Freezing point, °F (atmospheric)	Critical temp, °F	Condensing (head) pressure, at 86°F, psig[c]	Evaporation (suction) pressure at 5°F, psig[c]	Net refrig. effect (86 to 5°F) Btu per lb	Refrigerant flow (86 to 5°F) lb/(min)(ton)	Liquid refrig. flow (86 to 5°F) cu in./(min)(ton)	Compressor displacement (86 to 5°F) cfm per ton	Hp per ton	Type of compressor[d]	Compressor cooling
Ammonia...	NH_3	L–H	−28.0	−107.9	271.2	154.5	19.6	474.6	0.421	19.6	3.46	0.98	Rec	Water
Carbon dioxide....	CO_2	L, M	−109.3	−69.9	87.8	1028.3	316.8	55.5	3.61	167.0	0.96	1.83	Rec	Water
Dielene....	$C_2H_2Cl_2$	H	118.0	−70.0	470.0	15.8*	28.3*	114.3	1.75	38.3	111.2	0.97	Cen	Water
Ethyl chloride.....	C_2H_5Cl	M, H	54.5	−217.7	369.0	12.4	20.5*	142.3	1.41	44.4	22.62	0.91	Rot	Air
R 11......	CCl_3F	H	74.7	−168.0	388.4	3.6	24.0*	67.5	2.96	56.0	36.3	0.94	Rot, Cen	Water
R 12......	CCl_2F_2	M, H	−21.6	−247.0	233.0	93.2	11.8	51.1	3.91	83.8	5.81	1.01	Rec, Rot	W or A
R 21......	$CHCl_2F$	M, H	48.0	−211.0	353.3	16.5	19.3*	89.4	2.24	45.8	20.46	0.93	Rot	Air
R 22......	$CHClF_2$	U, L	−41.4	−256.0	204.8	159.8	28.3	69.3	2.89	68.1	3.60	1.02	Rec	Water
R 113.....	$C_2Cl_3F_3$	H	117.6	−31.0	417.4	13.9*	27.9*	53.7	3.73	66.4	100.9	0.98	Cen	Water
R 114.....	$C_2Cl_2F_4$	M, H	38.4	−137.0	294.3	22.0	16.1*	43.1	4.64	89.3	19.59	0.99	Rot	Air
Isobutane ..	$(CH_3)_3CH$	M, H	10.3	−229.0	272.7	44.8	3.3*	111.5	1.79	91.1	11.50	1.08	Rec, Rot	W or A
Methyl chloride..	CH_3Cl	M, H	−10.8	−144.0	289.4	80.0	6.5	150.2	1.33	40.9	5.95	1.02	Rec, Rot	W or A
Methylene chloride..	CH_2Cl_2	H	103.6	−142.0	421.0	8.4*	27.4*	134.6	1.49	30.9	74.3	0.97	Cen	Water
Sulfur dioxide....	SO_2	M, H	14.0	−103.9	314.8	51.8	5.9*	142.8	1.40	28.6	9.10	0.97	Rec, Rot	W or A
Water (40°F & 86°F)..	H_2O	H	212.0	32.0	706.1	28.7*	29.7*	1025.3	0.195	5.4	476.6	1.13	Cen, Jet	Water

Name of refrigerant	Safety data								
	Safety code group[e]	Low-side min test press. psig[f]	High-side min test press. psig[f]	Concentration in air to form flammable or explosive mix, % by volume[g]	Kills or seriously injures when in air: Exposure time, hr	Kills or seriously injures when in air: % by volume concentration	Kills or seriously injures when in air: Lb per 1,000 cu ft of air	Mix with lube oil?	Odor[h]
Ammonia...	2	150	300	16–25	½	0.5	0.25	No	Yes
Carbon dioxide....	1	1,000	1,500	N	½–1	29	33	No	No
Dielene....	2	30	30	5.6–11.4 D	1	2	5	Yes	Weak
Ethyl chloride.....	2	50	60	3.7–12 D	1	4	7	Yes	Yes
R 11......	1	30	30	N, D	2	10	35	Yes	No
R 12......	1	140	235	N, D	2	28	90	Yes	No
R 21......	1	40	70	N	½	10	27	Yes	No
R 22......	1	150	300	N, D	...	...	...	Yes	No
R 113.....	1	30	30	N, D	1	5	23	Yes	No
R 114.....	1	50	50	N, D	2	20	90	Yes	No
Isobutane ..	3	70	130	1.8–8.4	...	...	...	Yes	Weak
Methyl chloride..	2	120	210	8.1–17.2 D	2	2	2.5	Yes	Weak
Methylene chloride..	1	30	30	N, D	½	5	11	Yes	Weak
Sulfur dioxide....	2	85	170	N	5 min	0.7	1.2	No	Yes
Water (40°F & 86°F)..	...	...	...	N	...	...	...	No	No

NOTE: The data given are based on standard-ton conditions: evaporation at 5°F, condensation at 86°F.

[a] N, nitrogen; H, hydrogen; C, carbon; Cl, chlorine; F, fluorine; S, sulfur. [b] U, ultra-low: −130 to −75°F; L, low: −75 to 0°F; M, medium: 0 to 30°F; H, high: over 30°F. [c] Inches of mercury vacuum indicated by asterisk. [d] Rec, reciprocating; Cen, centrifugal; Rot, rotary. [e] ASHRAE Std 15-53: 1, used generally with minor limitations; 2, used generally with strict limitations; 3, prohibited in all but industrial occupancies (and commercial laboratories in unit systems containing not over 6 lb). See Standard for exact limitations for all refrigerants. [f] From ASHRAE Std 15-53. [g] N, nonflammable, except slightly flammable Freon-21; D, decomposition products from heat of flame are highly toxic. [h] Decomposition products of Freons, etc., have pungent odors.

Fig. 3-1. Better operation results from knowing the physical properties of the refrigerants with which one works.

were the first to develop and market this new synthetic refrigerant, the word "Freon" has become accepted as the name for all refrigerants of this type.

Q Are all Freons, or halocarbons, alike?

A See Fig. 3-2. No. The halocarbon family includes refrigerants with

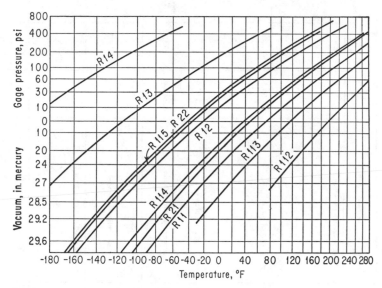

Fig. 3-2. Pressure-temperature relationships of Freon compounds.

atmospheric boiling points as low as −198.4°F (Refrigerant-14) to as high as 117.6°F (Refrigerant-113).

Q What is the refrigerant numbering system, and why is it important?

A See Fig. 3-3. A refrigerant numbering system has been proposed that assigns a specific number to each known refrigerant. It is necessary that some system be used because of the large number of new refrigerants with different properties being developed. These refrigerants should always be identified by name *and* number to avoid confusion.

> NOTE: "Carrene" is the trade name of refrigerants produced by Carrier Corporation; "Freon" by E. I. du Pont de Nemours & Company; "Genetron" by General Chemical, division of Allied Chemical Corporation; "Isotron" by Industrial Chemicals, Division of Pennsylvania Salt Manufacturing Co.; "Kulene" by the Eston Chemicals, Division of American Potash & Chemical Corporation;

New name	Old name	Formula	Chemical name	Boil temp., °F	Pressure, psig,* at		Freeze point, °F	Critical temp., °F
					5°F	86°F		
Refrigerant 11 Carrene 2 Freon 11 Genetron 11 Isotron 11 .... Frigen 11 Algeon 11	.... Carrene 2 Freon 11 Genetron 11 Arcton 9	CCl_3F	Trichloromonofluoromethane	74.8	24"	3.6	−168	338
Refrigerant 12 Freon 12 Genetron 12 Isotron 12 .... Frigen 12 Algeon 12	.... Freon 12 Genetron 12 Arcton 6	CCl_2F_2	Dichlorodifluoromethane	−21.6	12	93	−252	233
Refrigerant 13 Freon 13	.... Freon 13 Arcton 3	$CClF_3$	Monochlorotrifluoromethane	−114.6	177	†	−296	84
Refrigerant 13B1 Freon 13B1	.... Freon 13B1 Kulene 131	$CBrF_3$	Monobromotrifluoromethane	−73.6	63	247	−226	153
Refrigerant 14 Freon 14	.... Freon 14	CF_4	Carbontetrafluoride	−198.4	†	†	−312	−50
Refrigerant 21 Freon 21	.... Freon 21 Arcton 7	$CHCl_2F$	Dichloromonofluoromethane	48.1	19.2"	16	−211	353
Refrigerant 22 Freon 22 Genetron 22 Isotron 22 .... Frigen 22	.... Freon 22 Genetron 141 Arcton 4	$CHClF_2$	Monochlorodifluoromethane	−41.4	28	160	−256	205
Refrigerant 23 Freon 23	.... Freon 23	CHF_3	Trifluoromethane	−119.9				

Refrigerant 30.............	....	CH_2Cl_2	Methylene chloride or dichloromethane	105.2	27.6″	9.5″	−142	480
Carrene 30...............	Carrene 1							
Refrigerant 40.............	Methyl chloride	CH_3Cl	Methyl chloride	−10.8	6.5	80.0	−144	290
Refrigerant 50.............	Methane	CH_4	Methane	−259	†	†	−297	−116
Refrigerant 113............	....	$C_2Cl_3F_3$	Trichlorotrifluoroethane	117.6	27.9″	13.9″	−31.0	417
Carrene 113..............	Carrene 3							
Freon 113................	Freon 113							
Genetron 113.............	Genetron 226							
....	Arcton 63							
Refrigerant 114...........	....	$C_2Cl_2F_4$	Dichlorotetrafluoroethane	38.4	16.1″	22	−137	294
Freon 114................	Freon 114							
Refrigerant 114a..........	....	$C_2Cl_2F_4$	Dichlorotetrafluoroethane	38.5				
Genetron 114a............	Genetron 320							
Refrigerant 114B2.........	....	$C_2Br_2F_4$	Dibromotetrafluoroethane	117.5	27.8″	13.5	−167	418
Freon 114B2..............	Freon 114B2							
Refrigerant 115...........	....	C_2ClF_5	Monochloropentafluoroethane	−37.7	23	150	−159	176
Freon 115................	Freon 115							
Refrigerant 124a..........	....	C_2HClF_4	Monochlorotetrafluoroethane	14				
Freon 124a...............	Freon 124a							
Refrigerant 133a..........	....	$C_2H_2ClF_3$	Monochlorotrifluoroethane	43.0				
Freon 133a...............	Freon 133a							
Refrigerant 142b..........	....	$C_2H_3ClF_2$	Monochlorodifluoroethane	12.2				
Freon 142b...............	Freon 142b							
Refrigerant 152a..........	....	$C_2H_4F_2$	Difluoroethane	−12.4				
Freon 152a...............	Freon 152a							
Genetron 152a............	Genetron 100							

See page 30 for footnotes.

Fig. 3-3. Numbers label the ever-expanding list of refrigerants for quick and unmistakable identification. (*Continued on page 30.*)

New name	Old name	Formula	Chemical name	Boil temp., °F	Pressure, psig,* at 5°F	86°F	Freeze point, °F	Critical temp., °F
Refrigerant 160.........	Ethyl chloride	C₂H₅Cl	Ethyl chloride	54.0	20.5"	12.4	−218	369
Refrigerant 170.........	Ethane	C₂H₆	Ethane	−127.5	221.3	661.1	−278	90.1
Refrigerant 290.........	Propane	C₃H₈	Propane	−44.2	27.2	140.5	−310	202
Refrigerant C318.......	....	C₄H₈	Octafluorocyclobutane	21.1	9.6"	38		
Refrigerant 500.........	‡Carrene 500	....	‡	−28	16.4	113.4	−254	221
Refrigerant 600.........	Butane	C₄H₁₀	‡Butane	31.3	13.2"	26.9	−211	306
Refrigerant 601.........	Isobutane	C₄H₁₀	Isobutane	10.3	3.3"	44.8	−229	273
Refrigerant 717.........	Ammonia	NH₃	Ammonia	−28.0	19.6	154.5	−107.9	271
Refrigerant 718.........	Water	H₂O	Water	212	......	28.6"	32	706
Refrigerant 729.........	Air	....	Air	−318	......			−221
Refrigerant 744.........	Carbon dioxide	CO₂	Carbon dioxide	−109	317	1031	−69.9	87.8
Refrigerant 764.........	Sulfur dioxide	SO₂	Sulfur dioxide	14.0	5.9"	51.8	−104	315
Refrigerant 1150........	Ethylene	C₂H₄	Ethylene	−155	400	†	−272	48.8
Refrigerant 1270........	Propylene	C₃H₆	Propylene	−53.7	37	167	−301	196

*Except where given in inches of mercury vacuum.
†Above critical temperature.
‡Azeotropic mixture of Refrigerant 12 and Refrigerant 152a.

Fig. 3-3 (Continued)

"Ucon" by the Union Carbide Chemicals Corporation; "Arcton" by Imperial Chemicals Industries of England & Canada; and "Frigen" and "Algeon" produced in Germany and Argentina.

Q What is the effect of the halocarbon refrigerants on average lubricating oils?
A Most oils are completely miscible, or will mix with halocarbon refrigerants in all proportions. The warm refrigerant in the compressor discharge and receiver will carry large quantities of oil to the evaporator where it is concentrated due to the evaporation of the refrigerant. This tends to oil-log the evaporators and reduce their capacity. Care should be taken to remove as much oil from the refrigerant as possible before it reaches the evaporator. In flooded systems special provisions are made to return oil to an oil still or oil receiver.

Q What effect does a halocarbon refrigerant have on the pour point of oil?
A Most refrigerants of this type dilute the oil and reduce its pour point. This effect helps to remove oil from low-temperature evaporators, but it also reduces the lubricating effect of the oil.

Q Does ammonia have any effect on lubricants?
A Under normal operating conditions pure ammonia will not have any effect on a properly refined lubricating oil. Oil and ammonia will not mix. If allowed to come to rest, the oil will settle to the bottom of the vessel where it may be drained off. However, any oil that gets into the evaporators will also settle to the bottom. There it may present a removal problem, especially if the evaporator cannot be warmed up.

Q Why is it necessary to keep moisture from mixing with refrigerants?
A Anhydrous ammonia has no effect on metals, but the addition of small quantities of water causes ammonia to attack copper and its alloys. Moisture in ammonia systems can also cause sludging of lubricating oils. Moisture in systems using halocarbon refrigerants is much more critical. Very small quantities of water are enough to cause ice to form at the expansion valve and plug the system. A more serious effect is that water in contact with these refrigerants will create acids that attack all metal parts of the system. Great care should be taken to keep moisture out of the system, and adequate dryers should be kept in the system to remove any moisture that may get in.

Q Can one refrigerant replace another in a refrigerating system?
A Refrigerants in a system should not be changed until a thorough study of conditions and capacities has been made. Horsepower per ton, pounds circulated per ton, volumes per pound—all vary with different

refrigerants. They can create problems if not carefully studied. A system designed for one refrigerant will not operate effectively with another.

Q   Can refrigerants be mixed in a system?

A   Different refrigerants should never be mixed in an operating system. Mixtures of various halocarbon refrigerants are sometimes made for special purposes but should never be made in the field. The resulting mixture may have pressure-temperature characteristics entirely different from those of the components.

SUGGESTED READING

Reprints sold by *Power* magazine, costing $1.20 or less
  Refrigeration, 20 pages

Books
  Elonka, Steve, and Joseph F. Robinson: *Standard Plant Operator's Questions and Answers*, vols. I and II, McGraw-Hill Book Company, New York, 1959.
  Elonka, Steve: *Plant Operators' Manual*, rev. ed., McGraw-Hill Book Company, New York, 1965.
  Elonka, Steve, and Alonzo R. Parsons: *Standard Instrumentation Questions and Answers*, vols. I and II, McGraw-Hill Book Company, New York, 1962.
  ASHRAE Data Book, American Society of Heating, Refrigerating and Air-Conditioning Engineers.

4

COMPRESSORS

Q   How is a compressor used in a compression refrigeration system?

A   A compressor is used for one reason only—to save the expanded liquid so that it can be reused many times. If a bottle of ammonia was expanded into the cooling coils and discharged into the atmosphere, you would have the same refrigerating effect, but you would need a new bottle of refrigerant every time one emptied.

Q   Name four types or classifications of compressors.

A   Four broad classifications of mechanical compressors are (1) reciprocating, (2) centrifugal, (3) rotary sliding-vane, and (4) rotary screw compressor (rsc). Each classification may have several types of arrangement.

Q   Explain the basic differences between the four types.

A   See Fig. 4-1. (1) A reciprocating compressor consists of one or more piston and cylinder combinations. The piston moves in reciprocating motion to draw the suction gas into the cylinder on one stroke and to compress and discharge it to the condenser on the return stroke. (2) A centrifugal compressor has a single- or multistage high-speed impeller to set up enough centrifugal force within a circular casing to raise the pressure of the refrigerant gas to condensing level. (3) A rotary sliding-vane compressor is a positive displacement unit in that it traps a given volume of gas, compresses it, and ejects it from the machine. It usually has a rotor revolving off-center in a cylinder with sliding vanes forced against the cylinder wall. When space between any two vanes passes the suction port, the trapped volume is large. As this trapped volume moves around the cylinder, it becomes smaller and thus is compressed to its highest pressure as it is discharged from the cylinder through the discharge line. (4) A helical rotary screw compressor is another positive displacement unit. It was first used for refrigeration in the late 1950s, but because of its relative simplicity it is rapidly gaining favor. Basically, it consists of two mating helically grooved rotors, a male (lobes) and a female (grooves), in a stationary housing with suction and discharge ports. For sealing the lobes, oil (in most designs) is pumped through with the refrigerant. An axially movable sliding valve in the housing provides variable compressor displacement to satisfy the load requirements.

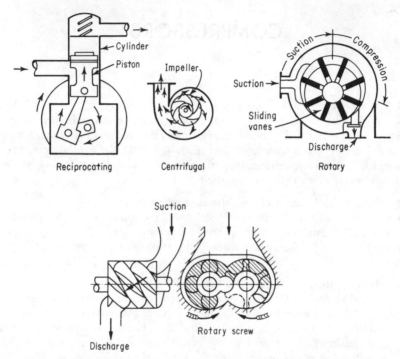

Fig. 4-1. Four types of refrigeration compressors in common use today.

Q   What is the historic background of refrigeration compressors?
A   Early refrigeration compressors were patterned after the steam engines in use at that time. These were large-cylinder horizontal compressors, usually with large-diameter pistons and speeds under 100 revolutions per minute (rpm). They were simple and rugged but required a large amount of floor space for a given capacity. To overcome this disadvantage, the

cylinders were placed in a vertical position supported by A frames. These cylinders were top-heavy and awkward to repair and to operate. Development of efficient high-speed electric motors lead to the evolution of smaller high-speed compressors.

Q What is an hda compressor?
A See Fig. 4-2. Hda is the abbreviation for horizontal double-acting compressor. This is a single- or two-cylinder compressor usually direct-connected to a steam engine or slow-speed synchronous motor. The pistons reciprocate in the cylinders in a horizontal plane. Suction and dis-

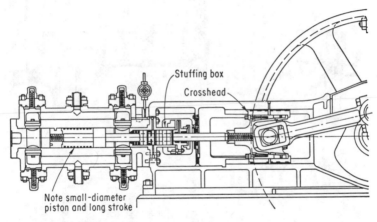

Fig. 4-2. Horizontal double-acting compressor (hda).

charge valves are placed in each end of the cylinder so pumping is done on each stroke in each direction. This is a very rugged type of compressor and is still manufactured for use in large installations.

Q What is a vsa compressor?
A See Fig. 4-3. Vsa is the abbreviation for vertical single-acting compressor. It is most commonly built in two-, three-, or four-cylinder combinations with medium rotative speeds, from 600 rpm in the smaller sizes to 300 rpm in the larger sizes. The crankshaft converts the rotary motion of the flywheel into a reciprocating motion to operate the pistons in a vertical plane. The pistons draw the suction gas into the cylinders on the down stroke and compress it to discharge pressure on the upward stroke.

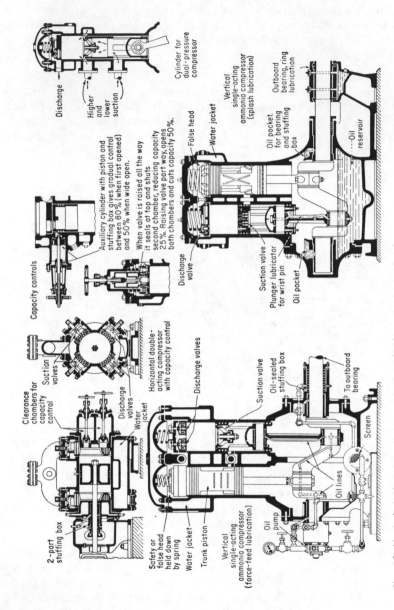

Discharge

Higher and lower suction

Cylinder for dual-pressure compressor

Auxiliary cylinder with piston and stuffing box gives gradual control between 80% (when first opened) and 50% when wide open.

When valve is raised all the way it seals at top and shuts second chamber, reducing capacity 25%. Raising valve part way, opens both chambers and cuts capacity 50%.

Capacity controls

False head

Water jacket

Vertical single-acting ammonia compressor (splash lubrication)

Oil pocket for bearing and stuffing box

Outboard bearing, ring lubrication

Oil reservoir

Discharge valve

Suction valve

Plunger lubricator for wrist pin

Oil pocket

Clearance chambers for capacity control

Suction valves

Discharge valves

Water jacket

Horizontal double-acting compressor with capacity control

Discharge valves

Suction valve

Oil-sealed stuffing box

To outboard bearing

2-part stuffing box

Safety or false head held down by spring

Water jacket

Trunk piston

Vertical single-acting ammonia compressor (force-feed lubrication)

Oil lines

Oil pump

Screen

Fig. 4-3. Vertical single-acting compressor (vsa) and other compressor details.

40

Q What is a V or VW compressor?

A See Fig. 4-4. V or VW compressors are types of single-acting compressors with cylinders and pistons arranged in a V, W, or VV pattern. Single compressors may have as many as 16 cylinders. This type is a high-speed compressor, often direct-connected to the shaft of the driving motor. Speeds to 1,750 rpm are common. Some experimental units have been operated at higher speeds.

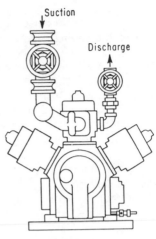

Q What is an angle compressor?

A See Fig. 4-5. An angle compressor is a compressor using one or more cylinders to pump the refrigerant vapor. It is driven by an internal combustion engine whose pistons and connecting rods are connected to the same crankshaft. One set of cylinders operates in a horizontal plane and the other in a vertical plane.

Fig. 4-4. V, W, or VV compressor.

Q What is a booster compressor?

A The conditions under which any given compressor operates determine whether a compressor is classified as a booster (see Chap. 2, Fig. 2-8). Although some specific designs of compressors are used only as

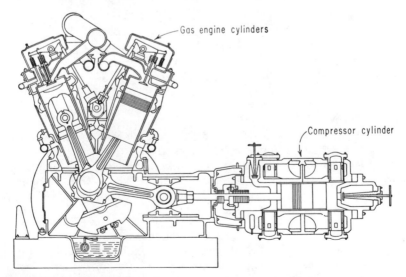

Fig. 4-5. Angle compressor.

boosters, any compressor may be used in booster service. In general, reciprocating machines that are designed for booster service have a cylinder diameter (bore) that is greater than the piston travel (stroke). For instance, $8\frac{3}{4}$ in. bore × 6 in. stroke, 13 × 9 in., and 15 × 10 in. are nearly always used as booster compressors. In industrial sizes all rotary compressors of the sliding vane type are recommended for booster service only.

Q   What is a hermetic compressor?
A   See Fig. 4-6. A hermetic compressor is a factory-sealed unit in which the compressor, either reciprocating or rotary, and motor are direct-connected and enclosed in a gastight housing. True hermetic compressors are factory-sealed and cannot be disassembled for repair in the field. They are limited to fractional horsepower units and up to $7\frac{1}{2}$ hp. Larger units, some to several hundred horsepower, are being built and have seen very successful operation. These units have the same advantages as hermetic compressors but should be classified as semihermetics, since they can be disassembled in the field.

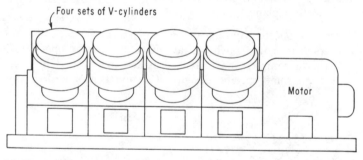

Fig. 4-6. Hermetically sealed reciprocating compressor.

Q   What is a Y compressor?
A   A Y compressor is a two-cylinder compressor with cylinders usually placed at a 45° angle from the vertical. It may have a motor built on the shaft. The cylinders may be the same size for single-stage operation, or one large and one small for two-stage operation. The Y type of compression is similar to the V type (Fig. 4-4), but Y compressors are usually built in the larger cylinder sizes.

Q   What is a dual- or multiple-effect compressor?
A   See Fig. 4-7. A dual-effect compressor handles suction gas at two different pressures. The regular suction ports connect with the lower-pressure suction, and the higher-pressure suction is connected to ports

in the cylinder wall above the piston location at the bottom of the stroke. During most of the stroke, low-pressure gas enters the cylinder through the suction valves in the piston. Near the bottom of the stroke, the ports of the higher-pressure suction are uncovered. The cylinder is thus filled with higher-pressure gas which compresses the lower-pressure gas and fills the remaining space. Both are then compressed to discharge pressure on the return stroke of the piston.

Q What is a stuffing box on a compressor?
A See Fig. 4-8. On a vsa compressor with the crankshaft extending through the crankcase wall, the refrigerant vapor must be contained within the crankcase when suction pressures are above atmospheric. Air must not enter the system when suction pressures are below atmospheric.

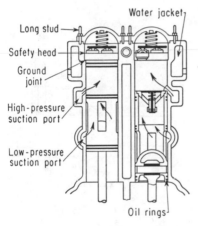

Fig. 4-7. Multiple-effect single-acting compressor.

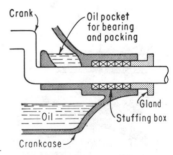

Fig. 4-8. Stuffing box holds shaft packing.

The stuffing box contains the shaft packing used to seal the crankcase against leakage. This is known as jam-type packing, since it is "stuffed" into the box around the shaft and held by a gland. On hda compressors the stuffing box contains similar packing, but it is placed around the piston rod.

Q What is an oil lantern ring on a compressor?
A See Fig. 4-9. The oil lantern ring is a spacer ring placed in a set of shaft or rod packing. It provides oil under pressure to lubricate the pack-

ing. On hda compressors where the packing is subjected to discharge pressure once on each stroke, the lantern ring is sometimes vented to the suction line. This relieves the high-pressure leakage past the first half of the rod packing. The main purpose of the lantern is to provide lubrication to the packing.

Q   What is a mechanical shaft seal and what is its purpose?
A   See Fig. 4-10. A mechanical shaft seal replaces the older jam-type packing for sealing around the shaft and crankcase. Instead of a number of turns of soft packing jammed into a stuffing box around the shaft, which eventually wears the shaft, one sealing ring and one mating ring are used. One is usually a hard steel, the other a carbon ring. Wear is confined to the two rings only, thus saving the shaft. Either O rings, gaskets (made of Neoprene), or metal bellows are used to prevent leakage between the rings and shaft or the housing. These seals need no adjustment once they are installed.

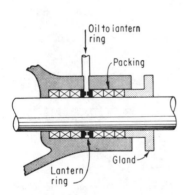

Fig. 4-9. Oil lantern ring introduces lubrication into packing set.

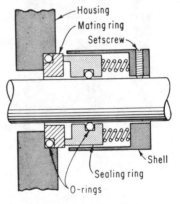

Fig. 4-10. Mechanical seal does not wear the shaft.

Q   How are refrigeration compressors lubricated?
A   See Chap. 10, "Lubrication." In general, vsa compressors are lubricated by splash or force feed, with lube oil supplied to some points by an external lubricator.

Q   What is the purpose of the water jacket on the cylinders of compressors?
A   Compressing the refrigerant vapor to a point where available condensing water or air can condense the refrigerant creates heat. Some of this heat is absorbed by the cylinder walls and if not removed can cause

lubrication problems. Most ammonia and CO_2 compressors have water jackets to assist in removing this heat. Freon compressors usually depend on air-cooled fins, since discharge temperatures are not as high as with ammonia and CO_2.

Q What is a safety head, and what is its purpose?

A See Fig. 4-11. In large vsa compressors the discharge valve is

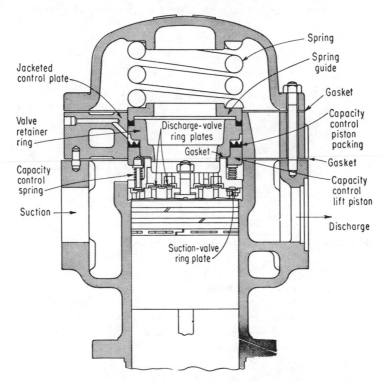

Fig. 4-11. Safety head prevents cylinder head from cracking.

usually mounted in a separate plate. This plate is held down by heavy springs that bear against the cylinder head. If liquid refrigerant or oil (which are noncompressible) enter the cylinder, the safety head will lift against the springs. This prevents serious damage to the compressors.

Q What is meant by compressor clearance?

A Compressor clearance is the amount of space between the top of the piston and the discharge valve plate when the piston is at the top of the stroke. This clearance should be as small as possible so that nearly all of the compressed refrigerant vapor will be forced past the discharge valve. Any high-pressure vapor left in the cylinder will expand during the suction stroke, partially filling the cylinder and thus reducing the cylinder volume.

Q   What is a clearance pocket?
A   See Fig. 4-12. A clearance pocket is a space designed into the head of the compressor to actually increase the effective clearance of the

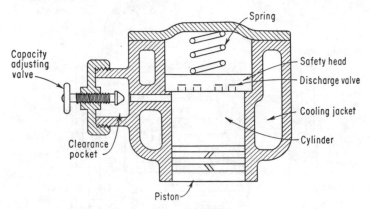

Fig. 4-12. Clearance pocket increases cylinder clearance.

machine for capacity control. When the clearance pocket valve is opened, either manually or automatically, the pocket is filled with high-pressure gas on each pressure stroke. On the suction stroke this high pressure vapor returns to the cylinder and must expand to the low pressure before more low-pressure vapor can enter the cylinder. This then reduces the capacity of the compressor without having to stop and start the machine.

Q   What is the difference between a clearance pocket and bypass capacity control?
A   See Fig. 4-13. As described above, a clearance pocket is a space at the top of the cylinder that increases the effective clearance of the compressor. It allows high-pressure vapor to re-expand into the cylinder on the suction stroke. The bypass capacity control is a valved port in the side of the cylinder wall. It allows the refrigerant vapor to escape back into the suction during the first part of the stroke. The percentage of capacity reduction is determined by the position of the port in relation to the length of the stroke. With the capacity-adjusting valve open,

cylinder volume below the valved port is bypassed. Some large vsa machines may have as many as three bypass ports on each cylinder to give several steps of capacity control.

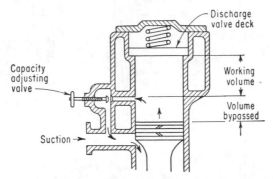

Fig. 4-13. Bypass capacity control helps control compressor output.

Q Name five other methods of reducing the capacity of refrigeration compressors.

A Besides clearance pockets and bypass controls, these five methods may be used to vary the capacity of compressors: (1) multispeed or wound-rotor motors may be used to vary the compressor speed; (2) the suction service valve may be partially closed so that the compressor is actually working at a lower suction pressure than that in the suction line; (3) discharge gas may be piped back into the suction of the compressor; (4) on centrifugal compressors prerotation vanes may be used; and (5) in modern VW compressors some cylinders may be unloaded.

Q What is an unloader?

A Unloading is usually done by a mechanical device that prevents the suction valves in given cylinders from closing. The unloading device may be operated by oil pressure from the lubricating system or by discharge gas pressure. Control can be by manually operated valves or by solenoid valves actuated from either pressure or temperature control. Many compressors of this type are arranged so that all unloaders are effective when the compressors start up. They remain unloaded until the compressor is up to operating speed or until the oil pressure reaches a predetermined point. This allows the use of normal starting-torque motors.

Q What are crossover valves?

A See Fig. 4-14. Crossover valves on older compressors are pipe connections (with valves) from the discharge line to the suction line. One leads from above the discharge line service valve to below the suction

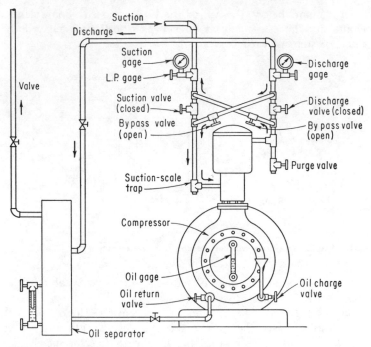

Fig. 4-14. Crossover hookup helps reverse flow through system.

service valve. The other leads from above the suction service valve to below the discharge service valve. If the main service valves are left closed and the crossover valves opened, it is possible to pump from the discharge side of the system into the suction side. In more modern vsa compressors, this whole arrangement is built into the compressor manifold (see Fig. 4-15).

Q What is a starting bypass, and what is its purpose?

A A starting bypass may be a separate valve between the discharge manifold and the suction manifold, or the manifold valves (Fig. 4-15) may be used as a starting bypass. The purpose of the starting bypass is to equalize the pressure between the discharge and suction sides of the compressor so that it may be started with no load on the motor. After the compressor is up to operating speed, the discharge valve is opened and the bypass is closed. A starting bypass is almost always used on compressors driven by synchronous motors.

Q What is a snifter valve?

A A snifter valve is a balanced valve between the suction manifold and the crankcase of some compressors. This valve maintains the crankcase at suction pressure and allows any oil returning with the suction gas to return to the crankcase. The balanced valve closes to prevent oil from entering the suction manifold in case of oil foaming.

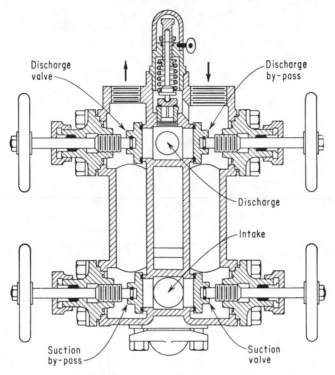

Fig. 4-15. Compressor manifold for ammonia compressor is a compact crossover system.

THE NEWEST COMPRESSOR, THE ROTARY SCREW

Q Describe the helical rotary screw compressor.

A Compressor designers have long dreamed of a machine combining the best characteristics of the positive displacement reciprocating unit and the continuous-flow rotary centrifugal machine, without the short-comings of either. The helical rotary screw (hrs) compressor, also known as the Lysholm type (Fig. 4-16), comes pretty close to fulfilling those wants. Today it is challenging the other designs.

The male rotor has four lobes, the female six. Thus the male rotor revolves 50 per cent faster. The female serves mainly as a rotating seal-ing member as the gas moves through the machine in an axial direction.

Normally, the inlet is at the top at one end, and the discharge outlet is at the bottom at the other end. At the inlet end, as the male lobe pulls out of the female lobe, the void draws inlet gas in through the inlet opening and a port in the inlet plate.

By the time the full length of the groove has drawn in a charge of inlet gas, the inlet port is cut off. This is about one-third of one revolution. A little later, a male lobe starts rolling down the female gully, starting at the inlet end. The opposite end of the gully is sealed at the discharge end by an end plate. As the male lobe squeezes the trapped gas within the gully, compression occurs.

The trapped gas at design pressure is forced through a port in the discharge plate as it is uncovered by a lobe. This same action occurs within subsequent female gullies.

Q   Explain how sealing is accomplished between rotor lobes.
A   Originally, the inventor used "dry" screws. The two screws (lobes) were kept apart (to prevent wearing) by timing gears on each shaft. Such machines had as great a temperature rise as any other displacement-type compressor. Also, there was slip (gas leakage) in the clearance between lobes and in the running clearances between rotors and cylinder, both around the circumference and at the end plates. To make the leakage a small proportion of the displacement, units were operated at very high rotation speeds, which made them noisy. The solution was the improved "wet screw" compressor. The liquid injected into the cylinder is lube oil, but its primary function is direct-contact cooling of the gas during the compression process. The oil also seals the running clearances and acts as a lubricant.

No other type of mechanical compressor can use large quantities of a cooling liquid within the compressor proper. This internal cooling would have the same advantages in a piston or centrifugal machine. But obviously it wouldn't work. Internal cooling means not only that there is a very low temperature rise due to compression, but also that very high compression ratios can be attained in a single stage without intercoolers.

Q   How is the capacity of the hrs compressor controlled?
A   If other types of compressors were controlled by throttling, the inlet pressure would be lowered. That would increase the compressor ratio and thus temperature rise, and perhaps cause surging.

Where power consumption is important, a slide valve is used at the inlet end of an hrs compressor, Fig. 4-16. This valve's function is to return to the inlet a variable portion of the full-rated displacement of inlet gas drawn in by the helical lobes. The slide valve can be controlled to provide infinite degrees of unloading, from full capacity to near zero.

Q   Are dry hrs compressors used for refrigeration?

A Yes, dry machines are often specified, but only for special applications. On the other hand, oil-flooded helical rotary compressors are designed to operate at high head pressures for all applications with all the usual refrigerants—R-12, R-22, R-502, and ammonia. Today, packaged hrs machines of 100 to 1,000 tons operating at 3,600 rpm are available.

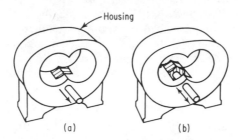

(a)   (b)

Slide valve for capacity control

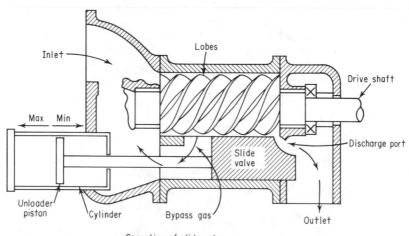

Operation of slide valve

Fig. 4-16. Slide valve of helical rotary-screw compressor is hydraulically actuated to regulate compressor capacity.

Q Explain the operation of the slide valve shown in Fig. 4-16.
A The capacity control slide valve is within the rotor housing. Axial movement of the valve shown is programmed by a solid-state, electronically initiated, hydraulically actuated control device. When the compressor is fully loaded, the slide valve is in the closed position (*a* in top sketch).

Unloading starts when the slide valve is moved back away from the valve stop (*b*) in Fig. 4-16. The movement of the valve creates an opening in the bottom of the rotor housing through which suction gas passes from the rotor housing back to the inlet port area before it has been compressed. Since no significant amount of work has been done on this returned gas, there are no appreciable losses. Reduced compressor capacity is obtained from the gas which is inside the inner part of the rotors and which is compressed in the ordinary manner. Capacity reduction down to 10 percent of full load is realized by progressive backward movement of the slide valve away from the valve stop. In principle, enlarging the opening in the rotor housing reduces compressor displacement.

Q   How is oil separated from refrigerant in hrs compressors?

A   Since a large amount of oil is injected at the inlet end, the discharge is a mixture of compressed gas and oil. The discharge goes to a discharge tank, where the bulk of the oil falls by gravity to a sump. What oil is left, in smaller particle or mist form, goes through a surface-type coalescer, which removes almost all of it. The gas leaving this separator system normally has less entrained oil than the gas from a piston compressor. Oil thus separated from the discharge gas stream is circulated through a heat exchanger, then back into the inlet of the compressor.

Each manufacturer has a differently designed oil separator. In one unit, a unique flow path pattern within the separator combines with an oversized mechanical demister.

Q   Explain how hrs packaged units are controlled.

A   The control center of these modern units can be supplied with all the operating controls necessary for automatic operation, plus numerous "fail-safe" controls and devices designed to protect the equipment under abnormal operating conditions. Major controls include (1) load limiter control, (2) anti-recycle timer, (3) low-pressure unloader controls, (4) oil pressure switch, (5) oil temperature control, (6) operating and freeze protection thermostats, and (7) hi-low–pressure switch.

Q   Sketch a two-stage refrigeration system using helical rotary screw compressors.

A   Figure 4-17 illustrates such an arrangement. Two-stage hrs units operating at an evaporating temperature of about −40°F use about 38 per cent less power than single-stage units of the same design. The arrangement shown is for a conventional refrigerating plant.

Intercooling between stages is not necessary if these compressors are used. Intercooling is replaced by a single liquid separator where flash gas produced is led to the suction side of the high-stage compressor.

The check valve in the low-stage suction line is for preventing backward rotation of the compressor due to a pressure differential between the high and low sides when the unit is shut down.

The hydraulic system which operates the sliding valve (for control) uses oil from the normal oil system of the compressor. Operating oil is controlled by solenoid valves.

While single-stage hrs units are best for plants with short running times, the most common use for hrs two-stage units is for industrial plants where continuous operation is required.

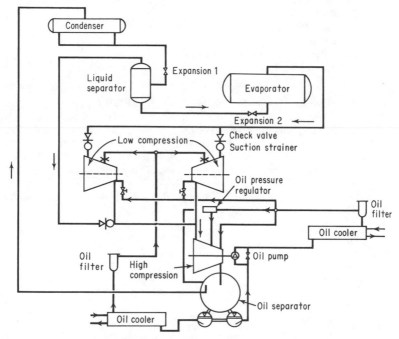

Fig. 4-17. Two-stage refrigeration system using hrs compressors for low temperatures needs no intercooling.

Q What is a hermetic compressor?
A A hermetic compressor has an electric motor and a compressor built into an integral housing, similar to those in home refrigerators. The motor and compressor have a common shaft and bearings. The motor is generally cooled by suction gas (refrigerant) passing through the windings of the electric motor, but it may be water-cooled.

These units eliminate the problems of motor mounting, coupling alignment, motor lubrication (due to lifetime lubrication), and refrigerant leakage at shaft seals, since the entire unit is hermetically sealed into one housing. Hermetic compressors may be (1) sealed (requiring factory service for repairs) or (2) accessible (permitting service on the job). Be sure you know which type you have before attempting repairs.

SUGGESTED READING

Reprints sold by *Power* magazine, costing $1.20 or less
  Refrigeration, 20 pages
  Gaskets, 20 pages
  Mechanical Packing, 24 pages
  Piston Rings, 24 pages
  Mechanical Seals, 24 pages
  Balancing Rotating Machinery, 24 pages
  Vibration Isolation, 16 pages

Books
  Elonka, Steve, and Joseph F. Robinson: *Standard Plant Operator's Questions and Answers*, vol. I, McGraw-Hill Book Company, New York, 1959.
  Elonka, Steve: *Plant Operators' Manual*, rev. ed., McGraw-Hill Book Company, New York, 1965.
  Elonka, Steve, and Alonzo R. Parsons: *Standard Instrumentation Questions and Answers*, vols. I and II, McGraw-Hill Book Company, New York, 1962.
  ASHRAE Data Book, American Society of Heating, Refrigerating and Air-Conditioning Engineers.

5

CONDENSERS

Q What is the purpose of the condenser in a refrigeration system?
A The condenser must convert to liquid all the refrigerant vapor delivered by the compressor.

Q How much heat does the condenser handle in relation to the heat absorbed in the evaporator?
A All the energy absorbed by the refrigeration system plus the heat equivalent of the mechanical energy required to operate the system must be disposed of by the condenser. For each 200 Btu per min absorbed by the evaporator, up to 300 Btu must be removed by the condenser. This depends on suction and discharge pressure and the type of refrigerant. The average system is designed for 250 Btu per min heat rejection for each 200 Btu refrigerating effect.

Q Name six common types of condensers used in refrigeration systems today.
A Common types of condensers include: (1) shell-and-tube, (2) shell-and-coil, (3) double-pipe, (4) atmospheric, (5) evaporative, and (6) air-cooled.

Q What are some points to consider in selecting a condenser for a refrigeration system?
A The condenser must have enough heat-transfer surface to condense the delivered vapor into a liquid state. This must be done at a reasonable operating pressure and temperature. The condenser must also have enough volume to store the vapor pumped by the compressor. Before condensing, the vapor occupies a definite volume; this volume can be decreased by increasing the pressure, but an increase in pressure causes an increase in the horsepower required to operate the system. If a condenser has sufficient surface, it usually has enough volume. Care should be taken when selecting finned-surface condensers. Finned surface may indicate sufficient area for heat rejection without providing enough volume. The condenser must also allow enough space for the

condensed liquid to separate from the vapor and drain to the liquid receiver.

Q   Where are air-cooled condensers used mostly?
A   Almost all fractional horsepower units are equipped with air-cooled condensers. They save by not needing water piping as do water-cooled condensers. In recent years the higher cost of water, increasing sewage disposal rates, and restrictions on the use of water have made air-cooled condensers in larger sizes popular.

Q   How much water is required per ton of refrigeration in a water-cooled condenser? How is this amount calculated?
A   The amount of water required will vary with the incoming temperature of the water and allowable rise in temperature through the condenser. The quantity of water required in gallons per minute (gpm) per ton is $250/[8.33(t_2 - t_1)]$ where $250 = $ Btu/(min) (ton) rejected in the condenser, $t_2 = $ water temperature leaving the condenser, $t_1 = $ water temperature entering the condenser, and $8.33 = $ pounds of water per gallon. If water is available at 80°F and allowed to gain enough heat to leave the condenser at 90°F, the quantity of cooling water required would be $250/[8.33(90 - 80)] = 250/(8.33 \times 10) = 3$ gpm of water required per ton capacity.

Q   What is the significance of the phrase "gallon-degrees per minute per ton"?
A   One gallon-degree per minute represents the heat required to raise the flow of one gallon of water per minute one degree Fahrenheit, or 8.33 Btu. If 250 Btu/(min)(ton) is required, then 30 gal = deg/(min)(ton) will be required. The gallons per minute flow of water required will be 30 gal-deg divided by the allowable temperature rise in the water through the condenser. Thus if the water may warm up 10°F, $30 \div 10 = 3$ gpm per ton; if only 6°F, $30 \div 6 = 5$ gpm per ton.

Q   Describe a shell-and-tube condenser.
A   A shell-and-tube condenser consists of a shell, tube sheets and tubes, water boxes, and refrigerant connections. In the smaller sizes the shell may be standard pipe, but welded shells are used in the larger sizes. Tube sheets, usually 1 or $1\frac{1}{4}$ in. thick, are welded to the shell and drilled to receive tubes. Tubes with ground or polished ends are inserted through their respective tube sheet holes and rolled or welded to provide a gastight joint. The refrigerant gas flows into the shell and around the tubes while water flows through the tubes.

Q   What is a vertical open shell-and-tube condenser, and what are some of its advantages?

A See Fig. 5-1. This type of condenser stands on end. Water is distributed over the entire head and enters each tube through a swirler to distribute the water evenly against the inside of the tube walls, where it flows downward by gravity. Some advantages are (1) large capacity

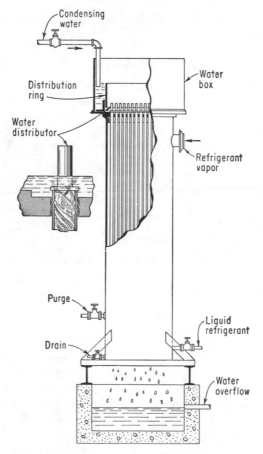

Fig. 5-1. Vertical shell-and-tube condenser has flow by gravity.

installed in small floor space, (2) low pumping heads required, (3) good gas and liquid separating space, (4) simple purging connections, (5) cleaning of tubes possible without stopping water flow (allowing water with some dirt to be used), and (6) ability to carry overloads by increasing water volume without increase in pump friction head

Q   What are the disadvantages of vertical shell-and-tube condensers?
A   (1) Inside installations have a tendency to "steam" in cold weather, (2) condenser must be mounted over an open water-collecting sump, and (3) relatively large quantities of water must be circulated because a single pass limits the water temperature rise.

Q   What is a horizontal closed shell-and-tube condenser, and what are its advantages?
A   See Fig. 5-2. A horizontal shell-and-tube condenser is equipped with enclosed water boxes and mounted in a horizontal position. Baffles

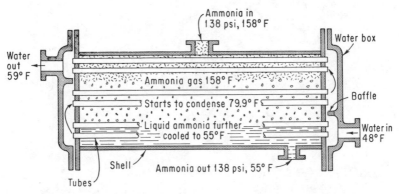

Fig. 5-2. Shell-and-tube condenser has one or more passes for cooling water through condenser; this one has three.

and gaskets in the enclosed water boxes cause the water to pass through the condenser several times. This allows the coolest water to enter at the bottom of the condenser and make several passes before leaving at the top. The result is a greater increase in water temperature with a corresponding decrease in water quantity. This type of condenser may be located at any point in a building. It should be installed so the heads may be easily removed for cleaning the tubes.

Q   What are some of the disadvantages of horizontal shell-and-tube condensers?
A   (1) They must be taken out of service for cleaning, (2) water pumping friction heads may be high, and (3) frequent cleaning may be needed to maintain efficiency.

Q   How do shell-and-coil condensers differ from shell-and-tube condensers?

A   Shell-and-coil condensers are usually used on smaller-tonnage low-pressure units. They consist of a shell that contains a coil for circulating the water. They do not have removable heads and the water side of the coil may be cleaned only with chemicals. In case of a coil leakage the entire coil bank must be replaced.

Q   What is a double-pipe water-cooled condenser?
A   See Fig 5-3. A double-pipe condenser has the condensing water tube inside the refrigerant tube. The refrigerant flows in the space

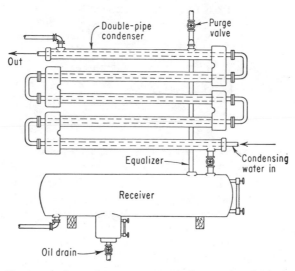

Fig. 5-3. Double-pipe condenser has one continuous loop, with refrigerant on outside, cooling water on inside.

between the tubes while the water is pumped through the inner tube. Water flows in the opposite direction to the refrigerant, with the coolest water in contact with the coolest refrigerant and the warmest water with the warmest refrigerant. Since copper tubing cannot be used with ammonia, this type of condenser is made up of steel pipe for ammonia, usually with a $1\frac{1}{4}$-in. pipe inside a 2-in. pipe.

Q   What are the disadvantages of double-pipe condensers?
A   Though the counterflow principle available with double-pipe condensers makes good use of the water available, the large number of joints and connections needed in large installations increases chances of leaks. These condensers are difficult to clean and do not provide enough space for separation of gas and liquid. For these reasons they

are not much used in large modern installations. Some small units are used in newer installations, but they must be cleaned chemically. In case of a leak, the entire unit must be replaced.

Q What is an atmospheric condenser?

A See Fig. 5-4. The atmospheric condenser was once very popular in large ammonia plants. It is built up of many lengths of pipe, usually 2-in.

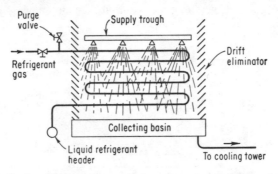

Fig. 5-4. Atmospheric condenser cools partly by evaporation; is little used today.

steel, with the ammonia vapor inside the pipe. The cooling water is distributed from a supply trough down over the outside of the pipes. Just as in a cooling tower (see Chap. 16), cooling is a combination of the evaporation of part of the water and the warming of the rest. Since water must flow downhill, some models introduced the hot refrigerant gas at the bottom for a counterflow effect. This caused problems in draining the condensed liquid from the units. Some were equipped with bleeders—small lines connected to the ends of each run to "bleed" the condensed refrigerant. Atmospheric condensers are seldom used today because of scale and algae problems and the large space required for a given capacity.

Q Describe an evaporative condenser.

A See Fig. 5-5. An evaporative condenser combines the functions of a cooling tower and a condenser. It consists of a casing enclosing a fan or blower section, water eliminators, refrigerant condensing coil, water pan, float valve, and spray pump outside the casing. The spray pump circulates water from the water pan at the bottom of the unit to the spray nozzles over the refrigerant coil. The fans pull or force air through the coil and through the water being sprayed over the coil. The heat from the refrigerant is transmitted through the metal coil to the water passing over the coil. The air removes the heat from the water by evaporating a portion of it. The eliminators prevent the water droplets from being carried out with the air.

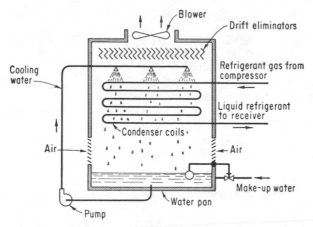

Fig. 5-5. Evaporative condenser is compact, highly efficient, very popular today.

Q What is a subcooling coil in an evaporative condenser?
A A subcooling coil is an auxiliary coil placed beneath the main coil. The liquid refrigerant is drained from the condenser to the receiver and is then piped through the subcooling coil on the way to the low-side equipment. This coil removes some heat from the liquid refrigerant and helps reduce the volume of flash gas.

Q What is a precooling or desuperheating coil in an evaporative condenser?
A A desuperheating coil is a separate coil used on some units to remove the heat of compression from the refrigerant gas before it enters the sprayed coil. This coil is designed to remove enough heat to reduce the temperature of the refrigerant to near the condensing temperature. This helps reduce scaling of the sprayed coil and reduces the relative humidity of the air leaving the unit.

Q What determines the capacity of an evaporative condenser?
A The capacity of an evaporative condenser depends on the amount of coil surface, the amount of air flowing over the coil, and the wet-bulb temperature of the air entering the unit. The total heat to be carried away is a function of the wet-bulb temperature. It represents the sum of the sensible and latent heat in the air at a given wet-bulb temperature. By determining the wet-bulb temperature of the air entering and leaving the condenser, the heat content at these two points may be

determined. The increase in total heat is due to the heat given up by the condensing refrigerant and would represent the capacity of the condenser. The lower the wet-bulb temperature of the entering air, the greater the capacity of the condenser (see Chap. 17).

Q   What is a bleedoff line on an evaporative condenser, and what purpose does it serve?
A   A bleedoff line is a small line leading from the pump discharge directly to the sewer. Since an evaporative condenser evaporates large quantities of water, the remaining water becomes increasingly concentrated in salts and solids. If these are allowed to build up, the problem of scale formation is increased. The bleedoff line assures that enough water will be pumped to the sewer to prevent the concentration from becoming too great. This idea is similar to a continuous blowoff line used on boilers.

Q   Which is the most satisfactory type of coil surface for evaporative condensers?
A   Both pipe (prime surface) and finned coils (extended surface) have been used in evaporative condensers for years. The bare pipe has some advantage in being easier to clean, but it is bulkier and heavier for a given capacity. The finned coil can be operated satisfactorily even under adverse water conditions with proper water treatment. It also has the advantage of providing enough capacity in subfreezing air temperatures when operated as a dry condenser.

Q   How should evaporative condensers be operated outdoors in subfreezing weather to prevent freezing the water in the unit?
A   Several methods provide safe operation of evaporative condensers in cold weather. (1) A separate water sump and pump may be located in the engine room or other heated space. The water from the condenser pan can flow to this indoor sump. (2) Electric or steam heaters controlled by a thermostat can be placed in the water pan. (3) Baffles or dampers can recirculate a portion of the heated discharge air through the condenser.

Q   How many refrigeration systems may be operated from a single evaporative condenser?
A   Evaporative condensers are now available with as many as 10 separate coil circuits. These units find use in supermarkets and similar operations where several compressor units are used on meat cases, vegetable displays, dairy boxes, etc. By using a condenser with a separate circuit for each unit, initial cost is reduced and each system is separated.

Q   What maintenance is required on evaporative condensers?

A Fan-shaft and motor bearings should be properly lubricated, fan belts should be checked periodically for tension and wear, and pump bearings should be lubricated. The water pan should be drained and cleaned at set periods, water spray nozzles should be checked and cleaned, rust and corrosion spots should be painted as required, and coils should be checked periodically for scale formations. Units having squirrel-cage type fans should be checked periodically for dirt built up in the fan blades, since this can cause unbalance and vibration to develop in the fans.

Q What causes loss of condensing capacity on partial loads in a system using more than one evaporative condenser?
A Because of the rapid condensation of refrigerant vapor, there is some pressure drop between the entering and leaving ends of the coil. If the fans on one or more condensers in a system are shut down to match the load, then no condensing takes place in those units, and there will be no pressure drop across the coil. The hot, high-pressure discharge gas will then flow through the coil of the inoperative unit into the liquid drain line and prevent the liquid from draining out of the coil of the operating unit. The trapped liquid then fills a portion of the coil and reduces its capacity. This condition can be avoided by (1) providing separate drain lines from each condenser to the receiver, (2) providing a trapped-liquid drain line of sufficient length to overcome the pressure drop, or (3) closing the inlet valve to a condenser whenever the unit is not in operation.

Q What are foul gases or noncondensables, and how do they affect condenser operation?
A Foul gases or noncondensables are a mixture or combination of hydrogen, nitrogen, water vapor, oil vapor, air, etc., inside the system. They circulate with the refrigerant but do no useful work and do not condense with the refrigerant. When these gases are present in the condenser, they have a tendency to collect on condensing surfaces and form a resistance to efficient heat transfer. This reduces the capacity of the condenser and raises the condensing pressure (see Chap. 14).

Q In a horizontal shell-and-tube condenser using a water-regulating valve, where should the valve be placed?
A The water-regulating valve should always be placed on the outlet side of the condenser so that all the tubes will be full of water at all times. If placed on the inlet side, siphoning or draining can sometimes leave some tubes partially full. This will lead to increased scaling and tube corrosion problems.

Q   Where should the inlet water connection be made on a multipass horizontal shell-and-tube condenser?

A   Water inlet connections on a multipass condenser should always be made at the bottom so that the coldest water is in contact with the coldest refrigerant. This will take advantage of the counterflow principle and will result in lower condensing pressure.

Q   What is the meaning of the term "fouling factor"?

A   Fouling factor refers to a factor used in calculating the over-all heat transfer through the tube walls of a condenser tube or other heat-transfer surface. It includes the sum of the heat-transfer rate of the layer of dirt and foreign material that builds up on the water side of the tube. If dirty water must be used or scale conditions are likely to occur, higher fouling factors should be used to ensure adequate condenser capacity under the worst expected conditions.

Q   How are various types of condensers cleaned of dirt and scale?

A   Shell-and-coil and certain types of double-pipe condensers can be cleaned only by circulating acids or other chemicals through them. This may be done on a continuous basis by adding a small amount of chemical to the water while the unit is in operation. Or, the unit may be taken out of service while a stronger solution is circulated for a short time. Evaporative condenser coils may be cleaned by scraping. Because the inner coils may be hard to reach, the most effective method is the use of acids or other chemicals. Shell-and-tube condensers are usually cleaned mechanically; that is, by a rotating cleaner (electric- or air-powered) or by pushing a properly sized scraper or brush through the tubes. Vertical shell-and-tube condensers may be cleaned while in operation, but horizontal types must be taken out of service and the heads removed. The

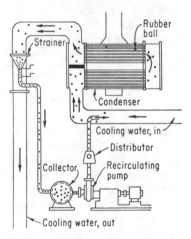

Fig. 5-6. Rubber balls pumped into circulating water provide continuous mechanical cleaning.

newest way to clean horizontal shell-and-tube condensers is to circulate a number of sponge-rubber balls through the tubes with an auxiliary pump while the condenser is in operation (see Fig. 5-6). A strainer on the outlet side returns the balls to the auxiliary pump suction for continued circulation. Proper water treatment will reduce the frequency of cleanings.

SUGGESTED READING

Reprints sold by *Power* magazine, costing $1.20 or less
> *Refrigeration*, 20 pages
> *Heat Exchangers*, 24 pages
> *Balancing Rotating Equipment*, 20 pages
> *Vibration Isolation*, 16 pages
> *Gaskets*, 20 pages
> *Piping*, 16 pages

Books
> Elonka, Steve, and Joseph F. Robinson: *Standard Plant Operator's Questions and Answers*, vols. I and II, McGraw-Hill Book Company, New York, 1959.
> Elonka, Steve: *Plant Operators' Manual*, rev. ed., McGraw-Hill Book Company, New York, 1965.
> Elonka, Steve, and Alonzo R. Parsons: *Standard Instrumentation Questions and Answers*, vols. I and II, McGraw-Hill Book Company, New York, 1962.
> *ASHRAE Data Book*, American Society of Heating, Refrigerating and Air-Conditioning Engineers.

6

EVAPORATORS

Q What part of the refrigeration system is the evaporator, and what is its importance?

A As the name indicates, the evaporator is that part of the system where the liquid refrigerant is evaporated (see Chap. 2, Fig. 2-2). It is sometimes called the cooling coil, unit cooler, freezer coil, liquid cooler, etc., but regardless of the name, that part of the system where the liquid refrigerant is changed into a vapor by the absorption of heat is an evaporator. Though an evaporator is sometimes a very simple apparatus, it is actually the most important part of the system. Any refrigeration system is designed, installed, and operated for the sole purpose of removing heat from some substance. Because this heat must be absorbed by the evaporator, the system's efficiency depends on proper design and operation of the evaporator.

Q What are the two main types of evaporators?

A All evaporators are classified according to the type of liquid feed into "dry" or "flooded" evaporators. A flooded evaporator (see Chap. 9, Fig. 9-13) is arranged with a tank or a surge drum (accumulator) located above the coil so that the inside of the evaporator is full, or flooded with refrigerant. A dry expansion coil (Fig. 6-1) is not "dry" but has a refrigerant control device that admits only enough liquid refrigerant to be completely evaporated by the time it reaches the outlet of the coil. All refrigerant leaves the coil in a dry state, that is, as dry vapor.

Q What are the three main requirements to be considered in design and selection of evaporators?

A (1) The evaporator must have enough surface to absorb the required heat load without excessive temperature difference between the refrigerant and the substance to be cooled. (2) The evaporator must provide sufficient space for the liquid refrigerant and also adequate space for the refrigerant vapor to separate from the liquid. (3) The evaporator must provide space for circulation of the refrigerant without excessive pressure drop between the inlet and outlet.

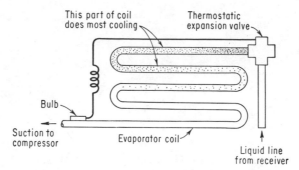

fig. 6-1. Prime-surface evaporator coil is bare pipe or tubing.

Q What is meant by prime-surface and extended-surface evaporators?
A Prime-surface evaporators are made from bare pipe or tubing (Fig. 6-1). Extended-surface evaporators or finned-coil evaporators (Fig. 6-2)

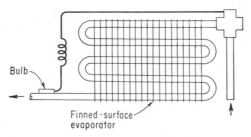

Fig. 6-2. Extended-surface evaporator coil has fins.

are made from bare pipe or tubing with extensions of the surface made by sheet-mctal plates, metal disks, or corrugations cast or machined into the surface of the tubing.

Q What are the advantages of extended-surface evaporators, and where are they used?
A One rule of good evaporator design is that the evaporator provide enough surface to absorb the heat load. Extended-surface or finned-coil evaporators present a much greater surface to the substance to be cooled than prime-surface or bare-pipe coils. Finned coils are usually restricted to use of cooling air or other gases.

Q What is a plate coil?

A See Fig. 6-3. A plate coil is an evaporator coil made of smooth metal plates with loops of pipe or tubing welded or soldered between

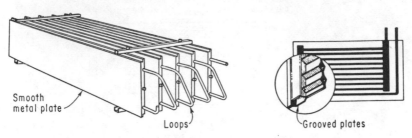

Fig. 6-3. Plate-coil evaporators come in smooth plate or in grooved plates.

the pairs of plates. They are also made of two embossed or grooved plates welded together so that the grooves form a channel for the refrigerant. They are most often used as shelf coils in freezers. The cold refrigerant is circulated through the channels, and the product to be frozen is placed between the plates.

Q Is the shell-and-tube water cooler also an evaporator?

A Yes, the shell-and-tube water cooler used in air-conditioning systems is a type of evaporator. This unit is similar in appearance to the shell-and-tube condenser and other heat exchangers. It is used to cool water, which is then circulated through the air-cooling units. In this design, water flows through the tubes of the cooler, while refrigerant liquid surrounding the outside of the tubes vaporizes as it absorbs heat from the water.

Q What is a headered coil?

A A headered coil has several lengths of pipe or tubing in parallel with the ends connected into a larger pipe, or header. This gives the advantage of a large amount of surface and a large volume of refrigerant without the disadvantage of long coils and high velocity.

Q What is an accumulator or surge drum, and how is it used with an evaporator?

A See Chap. 9, Fig. 9-13. An accumulator or surge drum is a vessel usually located near one end of and slightly above the evaporator coil. The liquid from the receiver is fed into the drum where the flash gas escapes immediately to the suction line. The remaining liquid flows by gravity to the evaporator coil where it is partially evaporated, and the mixture of gas and liquid returns to the accumulator. The refrigerant gas is carried away through the suction line, and the remaining liquid returns to the evaporator coil. A liquid-level device maintains liquid in the accumulator at a predetermined level.

Q   What is a direct-expansion evaporator?

A   See Fig. 6-4. A direct-expansion (dx) coil system is a direct method of refrigerating, in which the evaporator is in direct contact with the

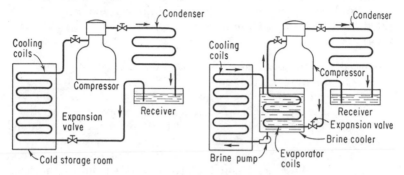

Fig. 6-4. Direct-expansion evaporator compared with the indirect one.

material or space refrigerated or is located in air circulation passages communicating with such spaces (see ASHRAE code). The evaporator of a direct system may include any heat-transfer vessel, such as pipe coils, shell-and-tube coolers, finned coils, or any apparatus in which a primary refrigerant, such as ammonia, Freon, or carbon dioxide, is circulated and evaporated for the purpose of cooling any material in direct contact with the opposite side of the heat-transfer surface. In contrast to this is the indirect system: the refrigerant is evaporated into the coils of the evaporator, which are in a brine tank. The brine secondary refrigerant is then circulated to the coils of the cold-storage box to do the cooling instead of the coils with the refrigerant inside. The distinction between a direct-expansion system and any other system is not in the size or shape of the heat-transfer apparatus, but in the process used to effect the heat transfer; by the latent heat process through evaporation of a primary refrigerant rather than the sensible heat process with a secondary refrigerant.

Q   Is a direct-expansion coil a flooded or a dry-expansion coil?

A   As explained above, any coil that places the refrigerant in direct contact with the heat-transfer surface that is absorbing heat from the substance to be cooled is classified as a direct-expansion coil. Therefore a direct-expansion coil may be either dry or flooded.

Q Why are ammonia coils that use thermal expansion valves usually bottom-fed, while the halocarbon refrigerants are usually top-fed?
A Any evaporator coil is more efficient if it is operated flooded, which means that most of the coil is filled with liquid refrigerant. All flooded systems feed liquid into the coil at the bottom and control the amount of wetted surface by controlling the height of the liquid. This can easily be done with ammonia because ammonia usually does not mix with oil (see Chap. 10, "Lubrication"). Halocarbon refrigerants are all miscible with oil and are top-fed so the oil may flow through the coil by gravity, assisted by the velocity of the refrigerant, and be returned to the compressor.

Q How does static head affect the operation of evaporator coils?
A The temperature of the evaporating refrigerant is determined by the total pressure at the point in question. A suction pressure gage will indicate the pressure of the vapor at the point where the gage line is connected. The pressure at the bottom of the coil may be several pounds higher due to the weight of the liquid refrigerant above it. Any increase in pressure creates an increase in the boiling point of the refrigerant. In a deep coil full of refrigerant the actual boiling point of the refrigerant may be several degrees higher at the bottom of the coil than at the top.

Q What is meant by the heat-transfer rate of an evaporator coil, and what are some of the factors involved?
A The heat-transfer rate of a coil is usually expressed in Btu per square foot of surface per hour per degree of temperature difference between the refrigerant and the substance being cooled. This is known as the heat-transfer coefficient U. The heat-transfer rate will vary depending on the (1) temperature difference, (2) method of feeding the coil, (3) substance being cooled, (4) velocity of the substance passing the coil surface and velocity of the refrigerant inside the coil, and (5) relative cleanliness of both surfaces.

Q What is a Baudelot cooler?
A A Baudelot cooler is a type of evaporator that cools liquids to near their freezing point. Early models had a series of pipes, one above the other. The liquid to be cooled flowed in a thin film down over the outside, and the refrigerant circulated within the pipes. Later models use stamped, corrugated stainless-steel plates (see Fig. 6-3), with the corrugation forming the channels for the refrigerant. The stainless-steel surface provides a sanitary, easily cleaned device. Also a continuous surface is provided, thus offering better control of liquid distribution. Any freezing that does occur has no effect on the plate.

Q What are some of the heat-transfer rates of liquid coolers?

A The heat-transfer coefficient U values may range as follows:

Flooded shell-and-tube water cooler.............	50–150
Flooded shell-and-tube brine cooler.............	30–100
Dry-expansion shell-and-tube water cooler,	
Freon in tubes, water in shell.................	50–115
Baudelot cooler, flooded, water.................	100–200
Baudelot cooler, dry expansion.................	60–150
Double-pipe cooler, water.....................	50–150
Double-pipe cooler, brine.....................	50–125
Shell-and-coil cooler.........................	10– 25
Spray-type shell-and-tube water cooler..........	150–250

Q What is the difference between flooded and dry-expansion shell and-tube brine or water coolers?

A The flooded cooler has the refrigerant in the shell area surrounding the tubes; the dry-expansion cooler has the refrigerant in the tubes, and the fluid to be cooled is pumped through the shell.

Q What are the advantages of each?

A Flooded coolers may be arranged for several passes of the fluid to be cooled, thus increasing the cooling range. High velocity in the tubes increases the heat-transfer rate. Dry expansion coolers have lower friction loss, simpler refrigerant charge, and tubes not subject to freezing and rupture. Over-all heat transfer is lower, but the actual installed cost of both coolers is about the same.

SUGGESTED READING

Reprints sold by *Power* magazine, costing $1.20 or less
  Refrigeration, 20 pages
  Air Conditioning, 16 pages

Books
  Elonka, Steve, and Joseph F. Robinson: *Standard Plant Operator's Questions and Answers*, vols. I and II, McGraw-Hill Book Company, New York, 1959.
  Elonka, Steve: *Plant Operators' Manual*, rev. ed., McGraw-Hill Book Company, New York, 1965.
  Elonka, Steve, and Alonzo R. Parsons: *Standard Instrumentation Questions and Answers*, vols. I and II, McGraw-Hill Book Company, New York, 1962.

7

REFRIGERANT CONTROLS

Q What are refrigerant controls?
A Refrigerant controls are devices used to control the flow of refrigerant at various points throughout the refrigeration cycle.

Q Name six points where refrigerant controls are used in a system, and describe briefly the basic function of each.
A The six points are: (1) Expansion valves, used to regulate the flow of refrigerant liquid to the evaporator. (2) Suction-line regulators, used to control the flow of refrigerant gas from the evaporator coil. This type of valve is also called a backpressure regulator or an evaporator-pressure regulator. (3) Hold-back valves, used to limit the flow of gas to the compressor to prevent surge or excessive loads from overloading the compressor motor. (4) Solenoid valves, used in liquid, suction, or discharge lines to interrupt the flow on demand from any one of several types of temperature- or pressure-sensing devices (see Chap. 8, "Electric Controls"). (5) Check valves, used to prevent the flow of gas from the condenser back to the compressor during off cycles. Check valves should also be installed in suction lines where higher pressures (from defrosting, etc.) could flow into another evaporator on the same circuit. (6) Reversing valves, used in heat pump applications or defrost cycles to change the flow of refrigerant.

Q What is an expansion valve, and what is its purpose in a system?
A "Expansion valve" is a standardized term used by the industry to describe any device that meters or regulates the flow of liquid refrigerant to an evaporator. The expansion valve also divides the high- from the low-pressure side of the system.

Q Name several types of expansion valves.
A There are six types: (1) capillary tubes, (2) hand expansion valves, (3) automatic expansion valves, (4) thermostatic expansion valves, (5) low-side floats, and (6) high-side floats.

Q   What is a capillary tube, and how does it qualify as an expansion valve?

A   Though the capillary tube is not a valve, it does answer the purpose of an expansion valve in household units and in some small commercial systems. It is a coil or length of fine tubing that has a very small orifice, usually 0.03 to 0.10 in. in diameter. The high pressure is dissipated in forcing the liquid through this small orifice, and a predetermined amount of liquid at a reduced pressure is allowed to flow to the evaporator. The capacity is determined by the diameter and length of tubing used; and, once installed, capacity cannot be changed except by changing the entire tube. The operation of a capillary tube is simple and foolproof; but absolute cleanliness of the refrigerant is necessary, since any foreign material can easily plug the small tube opening.

Q   What is a hand expansion valve, what are its advantages and disadvantages, and where is it used?

A   See Fig. 7-1. A hand expansion valve is a globe-type valve with a needle seat in the smaller sizes and a plug-type tapered seat in the larger sizes. The chief advantages of a hand expansion valve are its simplicity and low initial cost. Also, because of its simple construction there is very little that can get out of order. The main disadvantage is that an operator must be available at all times to make the necessary adjustments to meet changing load conditions. At one time it was the only expansion valve available, but it is rapidly being replaced by automatic devices. This valve is still used in large systems as a bypass valve (Fig. 7-2) around automatic control valves to allow operation in case of automatic valve failure and during repairs. Some flooded evaporator control systems also use a hand expansion valve, for throttling and for liquid control. These have a

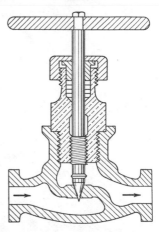

Fig. 7-1. Hand expansion valve with needle seat.

float switch and solenoid as the overriding control (see Chap. 8, "Electric Controls").

Q   Explain the automatic expansion valve.

A   The automatic expansion valve is a pressure-reducing device. It is actuated by the evaporator pressure which it keeps constant since the pressure of the refrigerant in the evaporator determines the evaporator temperature (see Fig. 7-3). It is a diaphragm- or bellows-operated valve, with the evaporator pressure acting on the lower side of the diaphragm and atmospheric pressure plus adjustable spring pressure acting on the upper side. As the compressor operates to remove the gas from the evaporator, reducing the pressure in the evaporator and under the dia-

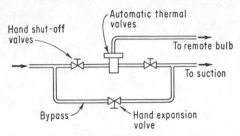

Fig. 7-2. Bypass around automatic thermal valve has hand expansion valve.

phragm, the adjusting spring pressure pushes the diaphragm down. This motion is transmitted through push rods (or by the needle valve stem) to the valve needle, opening it enough to allow more refrigerant to flow to the evaporator. As more liquid enters the evaporator, the pressure

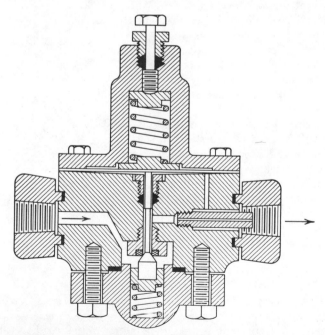

Fig. 7-3. Automatic expansion valve is diaphragm-operated.

increases, forcing the diaphragm upward and allowing the valve to close. A properly sized valve will pass enough liquid refrigerant to maintain constant temperature and pressure conditions.

Q Will an automatic expansion valve work well under changing load conditions? Why?

A An automatic expansion valve must be set to prevent overfeeding on low load conditions and therefore cannot feed enough on high load conditions. If the heat load drops off quickly, the evaporator pressure drops suddenly, opening the valve wide again in trying to raise evaporator pressure to the pressure setting of the valve. As a result the liquid refrigerant can flood back to the compressor and cause much damage. If the heat load increases suddenly, the evaporator pressure will increase rapidly, forcing the diaphragm up and allowing the valve to close. This starves the evaporator until the compressor can reduce the pressure and allow more refrigerant to pass into the evaporator. Hence, automatic expansion valves find their greatest use in systems with relatively constant loads and in systems with only one evaporator coil. One valve manufacturer uses the principle of capillary tubes with the automatic expansion valve by providing a long spiral path (similar to the thread on a bolt) for the refrigerant to flow through after passing the valve seat. This has the advantage of reducing the pressure drop across the valve seat, thus reducing seat erosion or "wire drawing" and the tendency to overfeed on low load conditions.

Q What is a thermostatic expansion valve?

A A thermostatic expansion valve is basically an automatic expansion valve with the added feature of a device to correct the feed rate of the valve so that it corresponds with the load on the evaporator.

Q Is the thermostatic expansion valve known by other names? .

A Yes. It is referred to as the thermal expansion valve and sometimes as the superheat valve. The term "superheat valve" is derived from the fact that the force needed to operate the valve is obtained from the superheat of the refrigerant gas in the evaporator coil.

Q Can the thermostatic expansion valve be used to control temperature?

A Not directly. The primary function of the thermostatic expansion valve is to meter the flow of refrigerant to the evaporator.

Q Describe a thermostatic expansion valve.

A Figure 7-4 shows that the thermostatic expansion valve is an automatic expansion valve with a temperature-sensitive element (also called "power element") replacing the adjusting screw and spring.

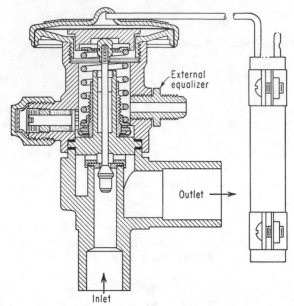

Fig. 7-4. Thermostatic expansion valve meters the flow of refrigerant to evaporator.

Q   Describe the operation of a thermostatic expansion valve.

A   See Fig. 7-5. With an evaporator operating at 37 psi on R-12 refrigerant, the saturation temperature is 40°F. As long as any liquid

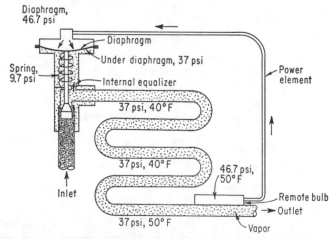

Fig. 7-5. Expansion of gas or liquid in remote bulb and power element keeps thermostatic expansion valve balanced to maintain flow.

remains in the coil, the temperature of the refrigerant-and-gas mixture will remain at 40°F. As the mixture progresses through the coil, heat is absorbed until at a point most of the way through the coil all the liquid is evaporated. As the refrigerant continues on through the coil, its temperature increases due to heat absorbed from the product or air being cooled. At a point near the coil's outlet, where the remote bulb is located, the refrigerant pressure remains at 37 psi, but the temperature increases to 50°F. The remote bulb, in close contact with the suction line, assumes the same temperature. Since the remote bulb and tube contain fluid with the same pressure-temperature characteristics as the refrigerant being used, the pressure within the remote bulb and the power element (tube) will correspond with the remote-bulb temperature of 50°F or 46.7 psi. This pressure of 46.7 psi is exerted on the upper side of the diaphragm. But this force is balanced by the evaporator pressure of 37 psi plus the spring pressure of 9.7 psi (corresponding to the required 10°F superheat). Under these conditions the valve is in balance. Any increase in heat load (or decrease in refrigerant) will increase the superheat and the pressure on top of the diaphragm, moving it to open the valve and admit more refrigerant to the evaporator. A decrease in load or increase in refrigerant flow will reduce the superheat and the pressure on top of the diaphragm, allowing the spring to move the valve to a closed position. Under operating conditions the valve achieves a balanced condition where the refrigerant flow balances the heat load.

Q   Are thermostatic expansion valves adjustable?

A   Most thermostatic expansion valves are set at the factory for a superheat of 10°F, designed so that an additional 4°F of superheat is required to move the valve to its fully open position. The superheat spring is adjustable within the range of 6 to 10°F. Generally speaking, the higher the superheat setting, the lower the evaporator capacity, since more of the coil is needed to create the necessary superheat to operate the valve. Superheat should be set to pass enough refrigerant to take greatest advantage of evaporator surface without allowing liquid refrigerant to reach the compressor.

Q   What are equalizers on thermostatic expansion valves, and when should each type be used?

A   Figure 7-5 illustrates diagrammatically an internal equalizer feature. An equalizer is a port or connection to transmit evaporator pressure to the underside of the diaphragm. Equalizers may be of two types: internal or external. Where the evaporator coil is fairly short and pressure drop through the coil is low, the use of an internal equalizer is best.

Usually, if pressure drop through a coil exceeds 5 psi on full load conditions, an external equalizer should be used to ensure full capacity of the coil.

Q   How does the use of external equalizers influence thermostatic expansion valve operation?

A   See Fig. 7-6. The first sketch represents an evaporator with a 10-psi pressure drop across the coil. With operating pressure of 37 lb and 40°F

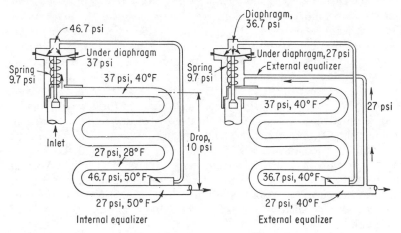

Fig. 7-6. Internal equalizer compared with external equalizer.

at the coil inlet, the force acts on the underside of the diaphragm to close the valve. Since there is a 10-psi pressure loss through the coil, the effective pressure at the remote bulb is 27 psi or 28°F. To achieve the required 10°F of superheat over coil entering conditions to balance the valve, it will be necessary to raise the coil outlet and remote bulb temperature to 50°F. This will need 22°F of superheat over coil leaving conditions and will require a greater percentage of coil surface. The second sketch (Fig. 7-6) illustrates a coil operating under identical conditions, but with an external equalizer. The coil entering and leaving conditions are the same, but the pressure exerted on the underside of the diaphragm (to close the valve) is the pressure at the coil outlet, transmitted through the external equalizer. The superheat now needed is 12°F. (The increase of 2°F in superheat is due to the change in pressure-temperature characteristics caused by the lower suction pressure of 27 lb.) Therefore, the capacity of an evaporator with a large pressure drop can be increased by using a thermostatic expansion valve with an external equalizer.

Q Where should the ends of the external equalizer be connected into the refrigerant circuit?

A Connect the valve end of the equalizer line into the fitting provided on the valve body itself. The downstream end of the line should be connected into the suction line at a point beyond any suction headers.

Q What is a multi-outlet or distributor type of thermostatic expansion valve? Where is it used?

A In the multi-outlet or distributor-type expansion valve the refrigerant is distributed in the valve itself through two or more openings. Each has a venturilike passage for maximum distribution efficiency with minimum pressure drop. Because the distribution takes place within the one valve with an outlet to each circuit, an equal homogenous mixture of gas and liquid refrigerant is assured for each evaporator circuit. This type of valve is used on any coil having more than one circuit through the coil.

Q Are remote bulbs and power assemblies on all thermostatic expansion valves the same?

A No. A thermostatic expansion valve must be selected for the specific refrigerant it will control and the type of service it will perform.

Q Name three types of remote bulb charges, and explain their differences.

A The three types of remote bulb charges are: (1) Liquid charge, where the amount of liquid in the remote bulb tubing and power head is greater than the combined volume of the power head and tubing. Under these conditions some liquid always remains in the remote bulb itself. The remote bulb and power head contain the same kind of refrigerant as that used in the system, which assures good control over a wide range of evaporator pressures. This valve operates well regardless of the location or temperature of the valve body itself. (2) Gas charge, which also contains the same refrigerant as that used in the system itself. However, the volume of liquid is limited so that at some predetermined temperature all the refrigerant in the power assembly and bulb will be evaporated. Any increase in remote-bulb temperature then only superheats the gas without greatly increasing the pressure. Since increased pressure would be needed to open the valve further to admit more refrigerant to the evaporator, this valve limits the amount of refrigerant and thus guards against flood back and motor overload. While this valve is not limited in application, it does work best on water chillers, air conditioners, etc., where evaporator temperatures range from 30 to 50°F. The valve itself *must* always be placed in a warmer location than that of the remote bulb. This prevents the gas charge from condensing in the valve assembly and making the valve inoperative. (3) Liquid cross charge, which contains a liquid whose pressure-temperature curve crosses the curve of the refrigerant used in the system in such a way that high superheat at high temperatures and low superheat at low temperatures are required. This valve is usually designed for specific temperatures and conditions. It is especially useful in low-temperature work.

Q   What general points should be considered in selecting a thermostatic expansion valve?

A   In order to select the right thermostatic expansion valve, the conditions under which it will operate must be known. These are (1) load, or tons of refrigeration, (2) size and type of inlet and outlet connections, (3) pressure differential across the valve at operating conditions, (4) possible need for an external equalizer, and (5) refrigerant used in the system. When all the conditions are known, select from a catalogue the valve that has the required capacity.

Q   How can location of the remote bulb affect the operation of thermostatic expansion valves?

A   See Fig. 7-7. If the remote bulb is located where liquid can become trapped, erratic operation of the valve will result. If the suction must rise

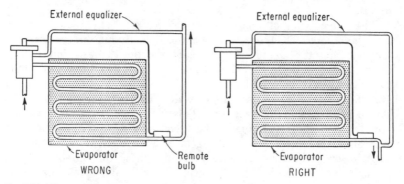

Fig. 7-7. Wrong and right way to hook up external equalizer.

from the evaporator, provide a trap downstream from the remote-bulb location (Fig. 7-8).

Q   What other precautions should be taken to ensure proper operation of thermostatic expansion valves?

A   A liquid line solenoid valve connected across the motor leads will prevent flooding of the evaporator during the off cycle in case of dirt in valve seats, corrosion, erosion, or worn seat surfaces. Strainers should always be installed ahead of each thermostatic expansion valve to prevent tiny particles of foreign material from damaging the precision valve parts.

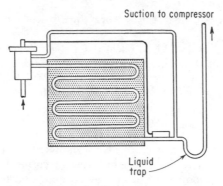

Fig. 7-8. Liquid trap is needed in suction line turning up.

Q What are the two types of float valves commonly used to control refrigerant flow to evaporators? Where are they found?

A The two types of float valves are low-side and high-side float valves. They are so named because of their location in the low-pressure or high-pressure portions of the refrigeration system.

Q How is a low-side float constructed, and how does it control refrigerant flow?

A See Fig. 7-9. A low-side float is essentially a hollow ball, pan, or inverted bucket connected through linkages and pivots to open or close

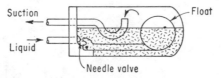

Fig. 7-9. Low-side float valve maintains level in evaporator.

a needle valve. It maintains a predetermined liquid level in an evaporator. As refrigerant is evaporated, the liquid level falls, lowering the float. The connecting linkage opens the valve to admit more refrigerant. As the liquid level then rises to the required point, the float is lifted and the valve closes.

Q What are some of the advantages of low-side float valves?

A A low-side float valve gives very good control by maintaining the proper refrigerant level regardless of load changes, compressor off cycles, load conditions, or other operating variables. Any number of evaporators may be operated on the same system, since each valve drains only the amount of refrigerant needed for its evaporator.

Q   Will all low-side float valves work equally well on all refrigerants? Explain.
A   No. Float valves should be selected for the specific refrigerant to be used because of the difference in density of refrigerants. A valve designed for one of the heavier refrigerants, such as R-12 or R-22, would have a smaller, heavier float ball than a valve designed for ammonia. Also, pressures in the system during defrosting, off cycles, etc., must be considered, since high pressures can collapse the float ball itself.

Q   What are some of the problems encountered with low-side floats?
A   The float ball may develop a leak due to corrosion or failure of a soldered or welded joint. The float may collapse due to high pressures. In either case the float would sink, causing the refrigerant to feed continuously and flood back to the compressor. The needle or seat, or both, may wear, causing it to leak continuously. The float ball may operate erratically because of the boiling action of the refrigerant. In such cases the float assembly is placed in a separate chamber.

Q   How is a high-side float valve constructed, and how does it control refrigerant flow?
A   See Fig. 7-10. A high-side float has the same elements as a low-side float: a hollow ball, linkages, and a needle valve. It differs from the low-

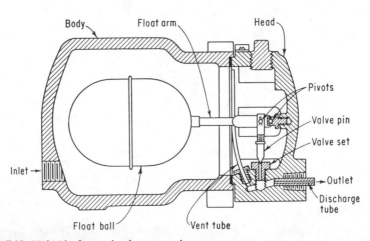

Fig. 7-10. High-side float valve has vent tube.

side float in that it is on the high-pressure side of the system and that a *rising* liquid level opens the valve. It is mounted below the condenser and passes the liquid refrigerant to the evaporator as rapidly as it is condensed, but it does not allow passage of uncondensed vapor. This requires that the major portion of the refrigerant charge be located in the evaporator.

Q Can more than one high-side float be used for refrigerant control in the same system? Explain.
A Since the high-side float normally passes all the liquid refrigerant reaching it, it is not practical to install this float in a multiple parallel evaporator system because no provision can be made for proper distribution of the refrigerant.

Q Why is a purge valve or vent tube necessary on a high-side float?
A If noncondensable gases are present in the system they will collect in the float chamber and prevent liquid from entering. This then starves the evaporator, fills the condenser, and makes the system inoperative.

Q What would occur if the float ball in a high-pressure float assembly became punctured or collapsed?
A If the float ball in a high-pressure float assembly did not rise with the rising liquid, the valve would not open. This would then trap all the liquid in the condenser and put the system out of operation.

Q Where may high-side floats be used?
A High-side floats find their greatest use in domestic and small commercial systems with a single compressor, evaporator, and condenser. They are also used in discharge oil traps to return oil to the system and in some types of automatic purgers.

Q What is a float switch and how may it be used to control refrigerant flow?
A See Fig. 7-11. A float switch is a liquid-level control device that makes use of a rising or falling float ball to actuate electrical contacts. The contacts in turn energize or deenergize one or more electrical circuits. They may be used (1) to maintain a liquid level in a flooded evaporator system, (2) to prevent loss of liquid seal in large receivers due to sudden load demands, (3) as sump or accumulator pump controls, or (4) as a device to actuate a warning circuit

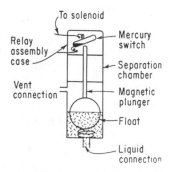

Fig. 7-11. Float switch has electrical contacts.

(light, bell, horn, etc.) on a predetermined high or low liquid level. They may be arranged to either open or close circuits on either a rising or a falling liquid level.

Q   What other components are usually used with a float switch in a system using flooded evaporator control?
A   See Fig. 7-12. A typical flooded evaporator system using a float switch also includes a solenoid liquid valve, strainer, and hand expansion

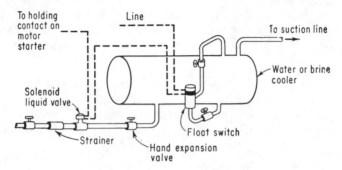

Fig. 7-12. These components are needed for hookup with float switch.

valve. In operation, the hand expansion valve is set to feed slightly more liquid than the evaporator requires for average operation. This helps to balance the system by regulating the refrigerant flow to prevent short cycling. It also reduces tendency of the evaporator to surge from sudden refrigerant flow when the liquid solenoid valve opens. As the liquid level in the evaporator rises due to the overfeeding of the hand expansion valve, the float switch opens or deenergizes the electrical circuit to the solenoid liquid valve, allowing it to close. When the liquid level drops to the float switch setting it closes the circuit, opening the solenoid liquid valve and admitting more liquid to the evaporator.

Q   How is a float switch used to prevent loss of liquid seal in a large receiver?
A   When used to prevent loss of liquid seal, a float switch is placed at the lowest allowable liquid level in the receiver. The electrical connections are so arranged that when the liquid level in the receiver falls to a predetermined point, the float switch deenergizes a circuit to the solenoid liquid valve, allowing it to close and thus preventing further flow of refrigerant from the receiver. As refrigerant is condensed and returned to the receiver, the liquid level rises; and at a predetermined point the float switch energizes the solenoid liquid valve circuit, opening the valve and allowing refrigerant to flow to the evaporator.

Q What precaution should be considered in any system using automatic control with float switches and solenoid valves?

A In general, control circuits to solenoid liquid valves should be connected to the compressor motor leads or to the holding contact on the motor starter to deenergize the solenoid circuit during the off cycle of the compressor. This will prevent the evaporator from filling up with solid liquid, which could cause serious damage due to surging when the compressor starts.

Q What piping and installation procedures must be observed to ensure proper operation of a float switch?

A The float switch must be mounted level and perpendicular and should be free from strains imposed by pipe connections. (Do not force connections!) Piping should be full size to allow free movement of refrigerant. Connect the vent or gas line from the side or top of the float chamber into the accumulator well above the maximum liquid level. Locate the float switch assembly so that the desired control level will fall within the "high" and "low" marks on the float chamber. Separate the float-chamber piping from other liquid piping. It is best to mount float switch assemblies with shutoff valves and union connections, so that the float switch may be removed for servicing or repair without pumping all the refrigerant from the system.

Q What is a solenoid valve, and where can it be used?

A A solenoid valve (sometimes called magnetic valve) is a shutoff valve that is actuated by an electromagnetic coil so designed that when the coil is energized, the magnetic field attracts an armature or plunger up into the core of the coil. By means of a valve stem or pin attached to the plunger, the valve is opened. Deenergizing the coil destroys the magnetic field, and the plunger falls of its own weight, closing the valve.

Q What are the two general types of solenoid valves, and how do they differ?

A See Fig. 7-13. Solenoid valves are divided into two general types:

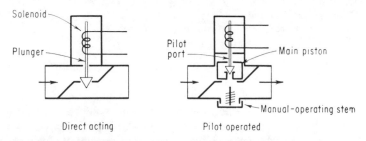

Fig. 7-13. Two types of solenoid valves: direct-acting and pilot-operated.

direct-acting and pilot-operated. A direct-acting solenoid valve is opened when the plunger pulls the valve pin away from the valve seat. To keep the solenoid coil to a reasonable size, the direct-acting valve is limited to smaller sizes. The pilot-operated solenoid valve does not open directly, but first opens a pilot port which releases the pressure on top of the valve pin, unbalancing the pressures acting on the valve seat. The higher upstream pressure then acts on the valve to open it. Solenoid valves are tight shutoff valves if kept in good repair, but because of being held closed by the pressure across the valve, they stop the flow in one direction only. If the pressure conditions reverse, that is, if pressure on the downstream side valve builds up to a higher pressure than the upstream side, the valve will open, permitting flow.

Q   Where are solenoid valves used in refrigeration systems?

A   A completely manually operated system may find no use for solenoid valves, but an automatic or semiautomatic plant may have solenoid valves in many locations, serving many functions. Since it is basically an electrically operated open-and-shut valve, it may be opened and closed by any one or a combination of sensing devices to maintain temperature, pressure, level, etc. Solenoid valves may be installed in main liquid lines and wired so as to stop the flow of refrigerant when the compressor stops. This prevents overflooding the evaporator during the off cycle. They may be installed in the liquid line and controlled by a pressure switch to shut off the liquid refrigerant flow when the suction pressure raises to a point that would overload the motor. They may be installed in liquid lines and controlled by a thermostat to shut off the liquid refrigerant flow when the desired low temperature has been reached. Solenoid valves are used in suction lines to isolate coils during defrosting operations, as bypass valves to reduce the capacity of a compressor, or as controls to operate capacity-reduction mechanisms. The solenoid valve may also be used as a bypass to allow starting of compressors under pressure with normal torque motors. A solenoid valve could be connected in the water-drain line of a condenser located outdoors and controlled to open at 34°F to drain the water and avoid freezing of the condenser. This valve may be used in almost any place where it is desirable or necessary to stop or start flow in a pipe subject to some type of remote device capable of opening and closing an electrical circuit.

Q   What two precautions should be observed when installing a solenoid valve?

A   (1) When installing a solenoid valve, make sure that the fluid flow is in the same direction as the arrow on the valve body. (2) Never wire a solenoid valve directly to the motor leads. The high current drawn to start the motor may drop the voltage enough to prevent the valve from opening. If the valve is to be opened as the motor starts, use a time-delay relay to delay opening the solenoid valve until motor is up to speed and voltage has returned to normal.

Q What are suction-line regulators, and in what two main classes do they fall?
A Suction-line regulators regulate the suction-line pressure to meet the requirements of the load or the capacity of the compressors. Two major classes are (1) the evaporator pressure regulator or backpressure control valve that controls the evaporator pressure and (2) the suction hold-back valve that controls the suction pressure at the compressor.

Q What is the purpose of an evaporator pressure regulator valve, and how does it work?
A The function of an evaporator pressure regulator is to prevent the evaporator pressure (and therefore the evaporator temperature) from falling below a predetermined point. In some cases it is used to adjust the evaporator pressure to meet changing load conditions. It may be

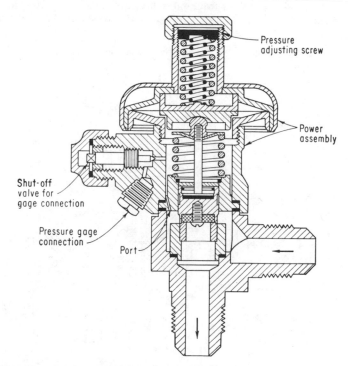

Fig. 7-14. Direct-acting evaporator pressure regulator.

used on a single evaporator, such as a water cooler, or several may be used on a single system to maintain desired pressures and temperatures in several different evaporators. In its simplest form (Fig. 7-14) it becomes a diaphragm-operated constant-pressure valve, with the upstream or evaporator pressure acting on the underside of the diaphragm. This pressure is balanced at the desired point by an adjustable spring pressure above the diaphragm. As the evaporator pressure increases, it overcomes the spring pressure, moving the valve stem in an upward direction and allowing refrigerant vapor to flow into the suction line. As the evaporator pressure decreases, the pressure spring closes the valve, maintaining the predetermined pressure.

Q　What is a pilot-operated evaporator pressure regulator?

A　See Fig. 7-15. A pilot-operated evaporator pressure regulator valve has either an internal or external pilot passage (but not both) to provide evaporator pressure to the pressure pilot. As evaporator pressure increases, it overcomes the spring pressure and opens the pilot port. This admits pressure into the space above the main piston, forcing it down and opening the main valve port. As evaporator pressure falls, the spring pressure in the pressure pilot closes the pilot port, the pressure above the main

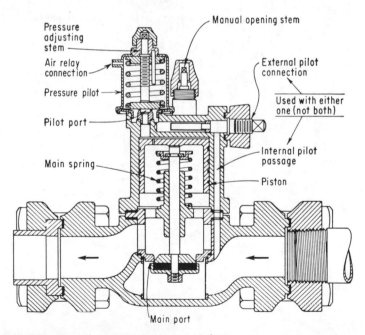

Fig. 7-15. Pilot-operated evaporator pressure regulator.

piston bleeds into the suction line, and the main spring closes the valve.

Q How can an evaporator pressure regulator valve be used to control temperatures?

A See Fig. 7-16. An evaporator pressure regulator valve can be used to maintain a specified temperature within very close limits by the use

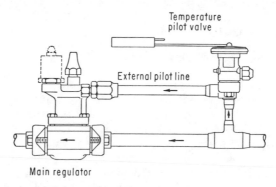

Fig. 7-16. Temperature pilot-operated regulator.

of a temperature-operated pilot piped to the external pilot connection of an evaporator pressure regulator valve. The remote bulb of the temperature-operated pilot is placed in the air stream or process flow to be controlled. As the temperature increases, pressure within the remote bulb increases, opening a port in the temperature pilot valve. This admits evaporator pressure through the external pilot line to the space above the piston in the main regulator, depressing the piston and opening the main valve. The evaporator pressure then drops to the pressure of the suction available, increasing the temperature difference and the capacity of the evaporator. As the temperature drops, pressure in the remote bulb decreases, closing the temperature pilot valve and reducing the flow from the evaporator to the main valve. As the pressure in the space above the main piston bleeds into the suction line, the main spring closes the main valve. Adding a solenoid valve in the external pilot line makes it possible to use this type of valve as a suction shutoff valve by properly controlling the solenoid.

Q What is an air-modulated evaporator pressure regulator?

A Refer to Fig. 7-15. Any evaporator pressure regulator valve with a pressure pilot spring housing can be converted to an air-modulated evaporator pressure regulator valve by adding a thermo-air relay or pneumatic control. The output pressure of the air relay (usually variable from

0 to 20 psig) is piped into the pressure pilot space. As the controlled temperature (remote bulb) increases, the thermo-air relay reduces the air pressure above the diaphragm, thereby reducing the total pressure (air pressure plus spring pressure) that must be overcome by the evaporator pressure to open the main valve. This effectively reduces the control point for the evaporator. As the controlled temperature decreases, the thermo-air relay increases the air pressure above the diaphragm, increasing the evaporator pressure in turn. This allows for a variation of up to 20 psi in the evaporator pressure.

Q   What is the purpose of a suction hold-back valve and how does it work?

A   A suction hold-back valve protects the compressor motor from overloads caused by suddenly increased loads due to defrosting, warm product, etc. In general it may be any one of several types of adjustable pressure diaphragm valves, with the control pressure being taken from the downstream or compressor side of the valve. As the compressor suction pressure increases, it acts on the diaphragm, closing the valve and limiting the flow of refrigerant gas to the compressor. As compressor suction pressure decreases, spring pressure opens the valve and admits more refrigerant gas to the compressor.

Q   Can a pilot-operated control act as a suction stop also?

A   See Fig. 7-17. Yes, if it is hooked up as shown in the sketch. Here the solenoid valve controls flow in the external pilot circuit. By properly

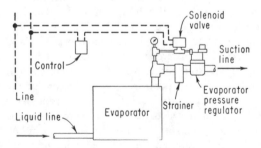

Fig. 7-17. Pilot-operated evaporator pressure regulator hooked up for stop duty.

actuating the solenoid valve, the evaporator pressure regulator can be closed when desired.

Q   Is a solenoid valve suitable for liquid-line shutoff service?

A   Yes, if it is designed for this job. Figure 7-18 shows a special valve suitable for liquid-line shutoff service on many different systems.

Q How does the valve in Fig. 7-18 work?

A The solenoid valve is connected between the liquid and gas lines and to the underside of the expansion-valve diaphragm. When solenoid pilot-control port is open, liquid passes from the high-pressure side

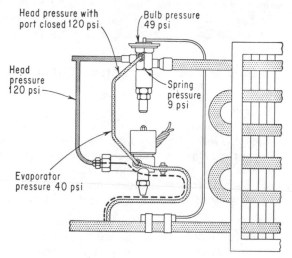

Fig. 7-18. Solenoid valve hooked up for liquid-line shutoff service on an evaporator.

through the port to the compressor suction line. If the port is closed, by actuating the solenoid valve, high pressure from the liquid line acts on the underside of the diaphragm to close the expansion valve, shutting off liquid flow to the evaporator.

Q What are the advantages of the valve hookup described above?

A Valves designed for this service are smaller and more economical than large-capacity solenoid valves. Usually only one size may be used on any capacity job. Lastly, they effect positive shutoff of the liquid line.

Q What precautions are needed in selecting a solenoid-valve power supply?

A See that power-supply voltage does not vary more than the amount the valve manufacturer allows. This is usually 10 per cent over or under rated valve voltage. Avoid excessive voltage drop, which may occur during starting, by wiring the valve into a separate circuit.

Q Are strainers needed with solenoid valves in refrigeration systems?

A Yes. Most manufacturers recommend a strainer with each valve even though there may be a master strainer somewhere else in the system.

Be sure that the strainer has enough capacity to prevent restriction of the flow through the solenoid valve. Always install a strainer ahead of the solenoid valve.

Q   What is the purpose of a manual opening stem on a solenoid valve?
A   It allows opening the valve by hand if the circuit to the coil is not in operation or if the valve coil burns out. The stem is used only in emergency and normally does not interfere with the valve operation.

TESTING REFRIGERATION SYSTEM CONTROLS

Q   Design a test panel for checking controls.
A   "Control equipment is only as good as the testing and maintenance it receives," has been said many times. Figure 7-19 shows a special test panel built by George Griffel, top-notch chief in a New York dairy plant. With it he checked the following (numbers refer to numbers on sketch): (1) temperature-controlled pressure-operated steam valves, (2) temperature controllers, (3) temperature recorders and thermometers, (4) pressure-controlled pressure-operated steam valves, and (5) pressure gages.

Q   How should temperature-controlled valves be tested?
A   A temperature-controlled valve is marked (1) on the sketch. And such a valve is installed with and controlled by a temperature controller (2). Notice that the controller has two pointers, marked A and B. Pointer A connects to a temperature bulb. Pointer B is set by hand to whatever temperature you want the controller to operate at. When in operation, the controller opens a valve to admit air to valve (1) when pointer A climbs to the setting of pointer B. Likewise, when A drops below B, the air supply to valve (1) is cut off, pressure is released from the diaphragm of valve (1), and the valve opens to allow full flow.

Both the valve and its air line are installed with unions so that replacement can be made fast. Whenever he learns that a valve is not opening or shutting off at the right temperature, here's what the maintenance man does.

First, he checks the setting of pointer B to make sure it hasn't been moved. Next he disconnects the air line from valve (1) and plugs it. Then he moves pointer B with the temperature-adjusting knob to see if the controller is operating and passing enough pressure to normally operate the diaphragm on valve (1). If the controller is okay, a new diaphragm valve is hooked in and the faulty valve is taken to the test panel and hooked up with a standard controller permanently mounted on the board. When the trouble is found and the valve repaired, it is returned to stock.

Q   How are temperature controllers tested?

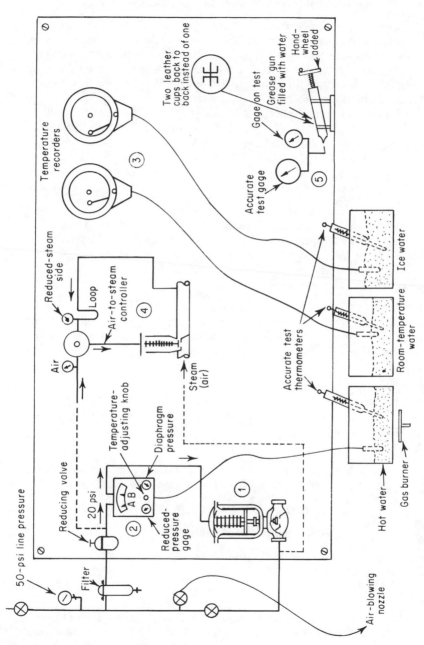

Fig. 7-19. Homemade panel for checking controls of refrigeration systems.

93

A   Turning the temperature-adjusting knob (with output air line disconnected and plugged) can usually show whether a temperature controller is working correctly. If it isn't, replace it immediately with a spare and take the faulty unit to the test panel. Replacing these temperature controllers takes more time than replacing the diaphragm valves because the thermal bulb has to be taken out along with the controller.

To check the controller at the test panel, put the thermal bulb in the hot water tank shown in Fig. 7-19 and hook its inlet air line to the reducing valve and its outlet air line to a previously tested diaphragm valve. Then it's easy to locate the trouble. Most troubles in the controller proper can be fixed, but if something is wrong with the thermal element it's cheaper to send the unit back to the factory for overhaul.

Q   How are thermometers and recorders checked?
A   Three different sources of temperature must be checked here. Use ice water, room-temperature water, and hot water. The temperature of the hot-water tank is regulated by a gas burner and can be varied. Place the bulb in each of these tanks in turn, then check the instrument reading against the test thermometer.

Q   How are pressure-operated valves tested?
A   Checking these pneumatically operated valves requires an air supply equal in pressure to the valve's operating pressure to hook up to the valve proper. Also, a variable-pressure air supply to hook up to diaphragm is needed.

Q   How are pressure and vacuum gages tested?
A   The homemade tester at (5) is nothing more than a grease gun fitted with a handwheel for fine adjustments. To keep air from leaking in during vacuum tests, replace the single leather cup with two cups back to back.

NOTE: The air-blowing nozzle shown at the lower left of the test panel is for blowing dust off instruments before working on them.

SUGGESTED READING

Reprints sold by *Power* magazine costing $1.20 or less
   Refrigeration, 20 pages
   Air Conditioning, 24 pages

Books
   Elonka, Steve, and Joseph F. Robinson: *Standard Plant Operator's Questions and Answers*, vols. I and II, McGraw-Hill Book Company, New York, 1959.

Elonka, Steve: *Plant Operators' Manual*, rev. ed., McGraw-Hill Book Company, New York, 1965.

Elonka, Steve, and Alonzo R. Parsons: *Standard Instrumentation Questions and Answers*, vols. I and II, McGraw-Hill Book Company, New York, 1962.

Elonka, Steve, and Julian L. Bernstein: *Standard Electronics Questions and Answers*, vols. I and II, McGraw-Hill Book Company, New York, 1964.

8

ELECTRIC CONTROLS

Q What are some of the electric controls used in modern refrigerating systems, and how do they differ from the refrigerant controls described in Chap. 7?

A Refrigerant controls are usually flow-control devices (valves, etc.) used to control the rate or amount of flow of some of the fluids in a system. These controls may be operated manually, by pressure, or by some outside influence. Electric controls are devices that make or break electrical circuits that can start or stop the entire system. They control the flow in some portion of the system, change the capacity of the compressors, provide automatic defrosting, and transfer liquid from one portion of the system to another. These controls perform many other functions in the automatic operation of a refrigerating plant. Electric controls include float switches, pressure switches, thermostats, relays, and their many combinations.

Q Name and describe the three types of pressure controllers.

A The three major types of pressure controllers, classified by their

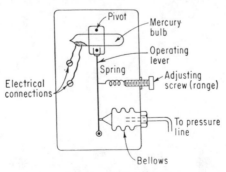

Fig. 8-1. Bellows pressure controller has a flexible bellows connected to pressure line.

power elements, are (1) bellows, (2) diaphragm, and (3) Bourdon tube. The bellows pressure controller (Fig. 8-1) consists of a flexible bellows connected to the pressure line to be controlled. The pressure in the line tends to expand the bellows but is opposed by a spring with an adjusting screw. The motion of the bellows is transmitted through linkages to open or close electrical contacts.

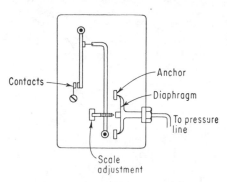

Fig. 8-2. Diaphragm controller has a flexible diaphragm.

The diaphragm controller (Fig. 8-2) employs a flexible diaphragm that performs the same function as the bellows. Since the movement of the diaphragm is somewhat limited, large leverages are needed to provide enough movement to actuate the switching mechanism.

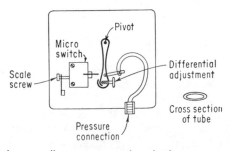

Fig. 8-3. Bourdon tube controller uses a curved oval tube.

The Bourdon tube controller (Fig. 8-3) consists of a curved oval tube securely anchored at one end and free to move toward or away from the switching mechanism at the other end. Again, linkages are provided to open or close electrical contacts or tilt a mercury bulb in relation to the pressure inside the tube.

Q In what way are temperature controllers (thermostats) similar to pressure controllers?

A   Many thermostats are designed around the same three basic power elements (bellows, diaphragm, and Bourdon tube) that are used in pressure controllers. Though they control from a variable temperature rather than pressure, the controlling force is the pressure created in the sensitive element by a fluid with definite pressure-temperature relationships.

Q   What is a bimetal thermostat?
A   See Fig. 8-4. A bimetal thermostat is a temperature-sensitive device made up of a thin duplex strip of two dissimilar metals with different

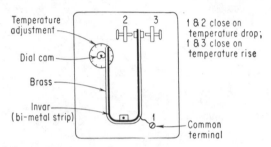

Fig. 8-4. Bimetal thermostat is a temperature-sensitive device.

thermal-expansion coefficients. As the temperature changes, the difference in expansion creates a deflection or bending action that opens or closes electrical contacts. The bimetal strip may be straight, in a U shape, or in the shape of a spiral for increased sensitivity.

Q   What is meant by the range and differential of a pressure or temperature controller?
A   See Fig. 8-1. The range of a controller is the average temperature or pressure at which the controller functions. It may be adjusted, within limits, by changing the tension of the range spring. The differential of any controller is the difference between the cutout and cutin points. Adjustment may also be at any given range, within the limits of the control. For example, a thermostat may be designed with a range of +20 to +120°F and a differential of 4°F. This means that at any point between +20 and +120°F the controller can be adjusted to operate at the 4°F differential. It can be set to (1) open a circuit to stop a compressor when the temperature falls to 40°F or (2) close a circuit to start the compressor when the temperature rises to 44°F.

Q   What is an adjustable differential thermostat?
A   An adjustable differential thermostat is a temperature control designed so that the operating or on-off differential may be adjusted within

the range of the control. One popular model has several ranges; from 0 to 70°F, to 160°F, and to 280°F, with differentials (on-off action) adjustable from ½ to 15°F.

Q   What is a remote-bulb thermostat?
A   A remote-bulb thermostat is a thermostat with the sensitive bulb connected to the control with a length of capillary tubing. This allows the control to be located at some distance from the substance being controlled (outside the room being refrigerated). These controls are usually equipped with about 5 ft of tubing, but additional lengths may be added.

Q   What do the following mean when included with the specifications for a control: SPST, SPDT, DPST, and DPDT?
A   SPST means single-pole single-throw; this control device has a single set of contacts and only opens and closes these contacts in response to temperature or pressure changes. SPDT means single-pole double-throw; this control device has a single movable contact that moves between two contact points, closing one circuit as the other opens. DPST means double-pole single-throw; this control device has two sets of contacts, may be wired to two different circuits or may open both lines of one circuit. It operates only to open or close circuits on temperature or pressure changes. DPDT means double-pole double-throw; this control is similar to SPDT, but it has two sets of contacts for controlling two circuits.

Q   What are the three types of remote-bulb thermostats according to bulb fill, and where is each used?
A   The three types of thermostats according to bulb fill are (1) the fade-out or limited fill (for low-temperature work), (2) the high-temperature fill, and (3) the cross-ambient fill. The limited fill is used where the sensing bulb is always at a lower temperature than the bellows. Because the bulb is always at a lower temperature, the liquid remains in the bulb and transmits proper pressure to the bellows. The amount of fill is such that at a given high temperature the liquid is all vaporized. A further increase in temperature will not create enough pressure to damage the control. The high-temperature fill is used where the bulb is always at a higher temperature than the bellows. In this type, though the bellows is completely filled with liquid, some liquid remains in the bulb to properly sense the correct temperature. Thus the bulb will always be the controlling point. The cross-ambient fill is used where the bulb may be either colder or warmer than the controller. The bulb is sized in relation to the bellows, so that if the bulb is colder than the controller the entire charge is contained within the bulb, which remains the control point. If the bulb is warmer than the bellows it fills with liquid, but enough liquid remains in the bulb to do the controlling.

Q   What is a low oil-pressure switch, and how is it used in a refrigeration system?

A   See Fig. 8-5. A low oil-pressure switch is a pressure controller whose contacts are wired into the compressor drive motor-circuit. When the oil

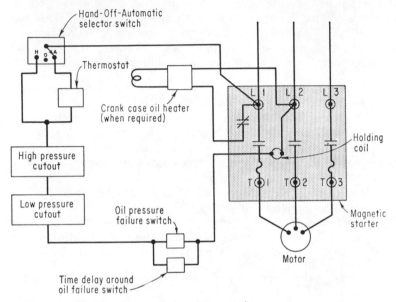

Fig. 8-5. Wiring diagram showing basic safety controls.

pressure developed by the compressor oil pump falls below the preset point, the controller opens the motor circuit to stop the compressor before damage due to lack of lubrication occurs. A time delay is usually connected around the main contacts. This makes the control inoperative on start-up until the pump has had time to build up the proper pressure.

Q   What are high-pressure cutouts and low-pressure cutouts, and what are the functions of each in a refrigeration system?

A   A high-pressure cutout is a pressure-control device used as a safety control on the discharge line of a compressor or a group of compressors. In case of condenser failure, or other operating conditions that cause the discharge pressure to rise above a set point, the high-pressure cutout opens the compressor motor control circuit to prevent further pressure increase. The control can also be wired to actuate an alarm circuit.

A low-pressure cutout is a similar type of pressure controller. It is so arranged that the contacts are opened when the pressure falls below a given point. It is sometimes used as a safety control to prevent the suction pressure from falling to a point where the compression ratio will be too great for the compressor design. It also prevents the suction pressure from falling to a point where other damage can occur from low temperature, such as the freezing-up of a water cooler. This cutout is often used as the control device to stop the compressor when pressure (and therefore temperature) conditions have been satisfied. These controls also have an adjustable range and differential.

Q   What is the difference between automatic and manual reset pressure controllers?

A   Manual reset pressure controllers are available in all pressure ranges and are so arranged that when the contacts are opened by a rise or fall of pressure beyond the set point, they must be returned manually to their closed position. Automatic reset controllers, as the name implies, automatically return to a closed position when the pressure returns to the operating area.

Q   What points must be considered in selecting pressure controllers to determine whether manual or automatic reset controllers should be used?

A   Generally speaking, if a pressure controller is to be used as a control device, an automatic reset controller should be used. If the controller is to be used as a safety device, a manual reset controller should be used. For example, a high-pressure safety cutout is designed to protect the system from excess pressures. If an automatic reset device were used, it would shut down the compressor on a dangerous increase in pressure and would restart the compressor as soon as the pressure had fallen below the set point. It would do this whether the cause of the trouble had been corrected or not. In case of condenser water failure this could lead to repeated cyclings of the compressor at the high-pressure point and possible damage to the compressor motor or drive. If a manual reset device is used, it becomes necessary for someone to manually reset the control. An inspection can be made to determine the cause of the overpressure.

Q   What is a floating control?

A   A pressure or temperature controller is sometimes installed to control a damper motor, motorized valve, capacity control, etc., in such a manner that a circuit is energized to open the damper on an increase in temperature, close the damper on a fall in temperature, but do neither when temperature remains between the set points. This type of controller employs an SPDT switch that energizes one circuit on an increase in temperature or energizes another circuit on a fall in temperature. Neither circuit is energized if the temperature remains constant. This is called a floating control.

LIBRARY
ROLLING PLAINS CAMPUS
TEXAS STATE TECHNICAL INSTITUTE

Q   What is a differential controller?

A   A differential controller is used where it is necessary to maintain a given pressure or temperature difference between two pipelines, processes, or spaces. These controllers, since they must sense pressure or temperature in two different locations, must have two bulbs, or sensing elements, and two bellows, or diaphragms (see Fig. 8-6). The bellows are connected by a rod so that movement of one bellows causes movement of the other. An adjustable spring is added to the low-pressure element to adjust the difference in pressures to be maintained so that the low pressure plus the spring pressure will equal the high pressure. An SPDT switch is usually furnished to energize proper circuits to operate the necessary valves, etc., and to maintain proper conditions.

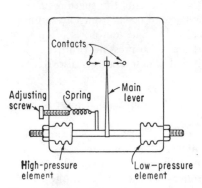

Fig. 8-6. Differential controller.

Q   What are float switches, and where are they used in refrigeration systems?

A   A float switch is an electrical-control device designed to open or close electrical circuits in response to the rise or fall of the liquid level in a vessel. It finds many uses in modern refrigeration systems: (1) to operate liquid solenoid valves to maintain liquid level in a suction accumulator for a coil, (2) to control operation of transfer pumps in suction liquid traps, (3) as operating controls for some types of liquid-return systems, (4) to control sump pumps, and (5) to actuate alarm lights or signal devices. It may be constructed so that the float ball actuates an electrical contact through mechanical linkage. The float may cause a rod to rise or fall inside a sealed tube, interrupting a magnetic field which then allows a permanent magnet to open or close an electrical circuit. Or it may have a similar function, interrupting an electronic field that in turn actuates a series of relays to control electrical circuits.

Q   What precautions should be observed in the installation and operation of any type of electric controls?

A   Normal precautions that apply to any electrical equipment should be observed. Controls should be solidly mounted, level, and plumb to ensure proper functioning. Wherever possible they should be mounted where they will be free from vibration and shock. Electrical connections should be made with a short length of flexible conduit. Care should be taken to keep electric controls clean and dry, and adjustments should be made only by qualified personnel.

Q What kind of safety controls are best for ammonia compressors?

A As most ammonia compressors today are motor-driven, installing high-pressure cutouts is important. Inspect them regularly and set them for pressures slightly above normal working conditions. The manual reset type is best because it calls the operator's attention to any trouble by having to be reset manually.

TROUBLESHOOTER'S GUIDE TO ELECTRICAL PROBLEMS

Trouble: Compressor Starts and Runs but Cuts Out on Overload

Probable cause	Symptoms	Recommended action
1. Low line voltage	1. Compressor cuts out on starter overloads	1. Measure line voltage. Contact power company to correct voltage
2. Additional current passing through starter overloads	2. Same as above	2. Check for incorrect wiring—fan motors, pumps, etc., connected to wrong side of starter overloads
3. Suction pressure high	3. Same as above	3. Check cutout current
4. Discharge pressure high	4. Same as above	4. Check cutout current
5. Starter overload heaters damaged or incorrect size	5. Same as above	5. Check current—replace starter overloads if damaged or wrong size
6. Running capacitor inoperative	6. Same as above	6. Check capacitor. Replace if inoperative
7. Stator partly shorted or grounded	7. Same as above	7. Check resistance, check for ground. Replace motor if necessary
8. Compressor tight	8. Same as above	8. Check resistance
9. Unbalanced line voltage (3 phase)	9. Same as above	9. Check voltage of each phase. If not equal, contact power company to correct condition of unbalance

Trouble: Starting Capacitors Burn Out

1. Rapid cycling of compressor	1. Repeated failure of starting capacitors	1. Replace starting capacitor. Correct rapid cycle condition. Check thermostat circuit for faulty thermostat or bad connections
2. Prolonged operation on starting winding	2. Same as above	2. Check for low line voltage. Contact power company to correct voltage

TROUBLESHOOTER'S GUIDE TO ELECTRICAL PROBLEMS (Continued)

Trouble: Starting Capacitors Burn Out (contd)

Probable cause	Symptoms	Recommended action
3. Relay contacts sticking	3. Same as above	3. Clean contacts or replace relay
4. Improper capacitor	4. Same as above	4. Check for proper capacitor rating

Trouble: Running Capacitors Burn Out

1. Excessive line voltage	1. Repeated failure of starting capacitors	1. Reduce line voltage so as not more than 10% under rated voltage of motor
2. High line voltage and light load	2. Same as above	2. Reduce line voltage so as not more than 10% over rated voltage of motor

Trouble: Relays Burn Out

1. Low line voltage	1. Repeated failure of relay	1. Increase voltage so as not more than 10% over rated voltage of motor
2. Excessive line voltage	2. Same as above	2. Reduce voltage so as not more than 10% over rated voltage of motor
3. Incorrect running capacitor	3. Same as above	3. Replace running capacitor with correct capacitance and voltage
4. Rapid cycling of compressor	4. Same as above	4. Same as above
5. Relay vibrating	5. Same as above	5. Secure relay on mounting
6. Incorrect relay	6. Same as above	6. Replace with correct relay

Trouble: Compressor Will Not Start—Hums, but Cuts Out on Starter Overload

1. Improperly wired	1. Compressor will not run— hums	1. Check wiring against wiring diagram
2. Low line voltage	2. Same as above	2. Check mainline voltage— determine location of voltage drop
3. Open starting capacitor	3. Same as above	3. Replace starting capacitor
4. Relay contacts not closing	4. Same as above	4. Check by operating manually. Replace inoperative relay
5. Open circuit in starting winding	5. Same as above	5. Check wires to stator. If wires OK, replace motor
6. Stator winding grounded	6. Same as above	6. Check wiring to motor. If wiring OK, replace motor

TROUBLESHOOTER'S GUIDE TO ELECTRICAL PROBLEMS (Continued)

Trouble: Compressor Will Not Start—Hums, but Cuts Out on Starter Overload (*contd*)

Probable cause	Symptoms	Recommended action
7. High discharge pressure	7. Same as above	7. Eliminate cause of high discharge pressure. Make sure compressor discharge shutoff valve is open
8. Tight compressor	8. Same as above	8. Check oil level. Repair compressor if damaged

Trouble: Compressor Starts, but Motor Will Not Get Off Starting Winding

Probable cause	Symptoms	Recommended action
1. Low line voltage	1. Motor runs continuously on starting winding	1. Measure line voltage. Contact power company to correct voltage
2. Improper wiring	2. Same as above	2. Check wiring against wiring diagram
3. Defective relay	3. Same as above	3. Check operation manually, replace relay if inoperative
4. Running capacitor shorted	4. Same as above	4. Check running capacitor
5. Starting and running windings shorted	5. Same as above	5. Check resistances. Replace motor if burned out
6. Starting capacitor open	6. Same as above	6. Replace starting capacitor
7. High discharge pressure	7. Same as above	7. Check cutout current
8. Tight compressor	8. Same as above	8. Check cutout current

Trouble: Compressor Will Not Run

Probable cause	Symptoms	Recommended action
1. Open line circuit	1. Compressor will not run—no hum	1. Check wiring and fuses
2. Overload cut out	2. Same as above	2. Wait for reset, then check current
3. Control contacts open	3. Same as above	3. Check controls, check pressures
4. Open circuit in motor	4. Same as above	4. Replace motor

SUGGESTED READING

Reprints sold by *Power* magazine, costing $1.20 or less
  Electrical Protection, 32 pages

Books
  Elonka, Steve, and Joseph F. Robinson: *Standard Plant Operator's Questions and Answers,* vols. I and II, McGraw-Hill Book Company, New York, 1959.

Elonka, Steve, and Alonzo R. Parsons: *Standard Instrumentation Questions and Answers*, vols. I and II, McGraw-Hill Book Company, New York, 1962.

Elonka, Steve, and Julian L. Bernstein: *Standard Electronics Questions and Answers*, vols. I and II, McGraw-Hill Book Company, New York, 1964.

9

PIPING AND FITTINGS

Q What effect does velocity have on pressure drop in piping?
A The higher the velocity through piping, the greater the pressure drop. All lines having flow through them have a pressure drop. The amount of drop depends on the size of the pipe and the length and number of bends in the system. If piping is sized incorrectly, or if many bends are placed into a long line that is undersized, the pressure drop can be so great that the system will not operate properly.

Q What is the usual pressure drop from the receiver to the expansion valve?
A There is usually a 5 psi pressure drop or less from the receiver to the expansion valve. This drop should never be more than 10 psi. To prevent vaporization of the refrigerant ahead of the expansion valve, it is best to maintain a velocity of 100 to 250 feet per minute (fpm).

Q How does the pressure drop in the suction line between the evaporator and compressor affect operation?
A Pressure drop in the suction line between the evaporator and compressor results in a lower pressure being maintained inside the compressor than in the evaporator.

Q How does the pressure drop between compressor and condenser affect operation?
A The pressure drop between compressor and condenser means that a higher pressure must be kept inside the compressor during discharge than is kept in the condenser.

Q What kind of piping is used in R-12 plants?
A In early installations of R-12 plants seamless steel tubing with welded, flanged, or gasketed joints was used. This was the common practice in plants using ammonia, carbon dioxide, sulfur dioxide, etc. But because of leaks caused by the solvent action and searching qualities of R-12, the present practice is to use seamless copper tubing with soldered or flanged joints.

Q   What kind of piping joints are used in halocarbon piping connections?

A   See Fig. 9-1. Where flanged joints are required, the tongue-and-groove type are used with compressed asbestos or metallic gaskets. Gas-

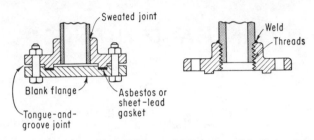

Fig. 9-1. Tongue-and-groove, welded, and sweated joints are used for Freon piping.

kets impregnated with oil, graphite, etc., are unsatisfactory because Freon-12 dissolves these substances and leaks result. Threaded pipe joints should also be avoided, but if it is necessary to use this kind of joint, it should be seal-welded or soldered (sweated) after assembling.

Q   Why is a liquid receiver necessary, and what connections does it have?

A   See Fig. 9-2. The liquid receiver serves the following functions: (1) stores unused refrigerant returning from the condenser, (2) stores re-

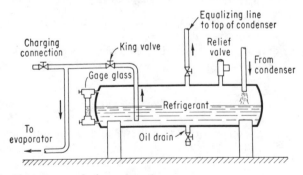

Fig. 9-2. Liquid receiver needs these connections.

frigerant to be evaporated by the expansion valve, (3) stores excess refrigerant in the system, and (4) provides a place to store refrigerant when pumping out the evaporator during maintenance operations. The receiver must have a return line from the condenser, a relief valve, and an equalizing line to the top of the condenser. This vent line equalizes the pressure in the condenser and receiver so that condensed refrigerant will flow from the condenser to the receiver. The liquid line extends into the receiver to within a few inches from the bottom, so it does not pick up dirt or oil settling out. A gage glass shows the liquid level at all times. An oil drain at the bottom of the receiver is used to remove oil that is carried along with the refrigerant.

Q Show how to install a discharge line where the compressor load changes considerably and the condenser is high above the compressor.
A See Fig. 9-3. The double riser shown in this sketch is the answer.

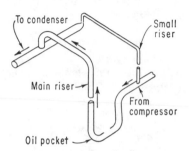

Fig. 9-3. Use double risers when gas velocity goes below 1,500 fpm at low capacity.

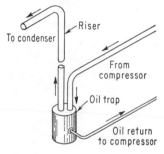

Fig. 9-4. Oil trap may be used instead of a double riser. Trap returns the oil.

Gas passes through the small riser at 1,500 fpm, the velocity needed to carry oil. At maximum load gas flows through the pocket, carrying oil

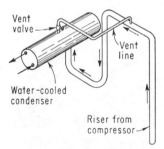

Fig. 9-5. Loop hot-gas line before entering condenser. Vent cuts siphoning action.

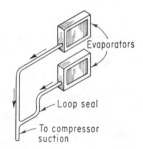

Fig. 9-6. Connect suction line like this when evaporators are above the compressor.

with it at 1,500 fpm. Another way to get the same results is to use an oil trap (Fig. 9-4). But the riser must be sized for normal pressure drop at maximum load. The oil goes directly from the trap to the compressor.

Q   Explain how to install a discharge line where the condenser is above the compressor and has a bottom inlet.

A   See Fig. 9-5. Loop the discharge line so it forms a seal. The top of the loop should be 6 ft above the condenser liquid level. If this much space isn't available, connect a vent line as shown in the sketch. Be sure to close the vent valve after pumping down the system.

Q   Show how to make a suction piping hookup where two or more evaporators are above or at the same level as the compressor.

A   See Fig. 9-6. Here the lower suction line from the evaporator must have a loop. This hookup prevents oil from the top evaporator from collecting in the lower one while the system is shut down.

Q   Explain how to hook up suction lines where two or more evaporators are below the compressor.

A   See Fig. 9-7. Where two or more evaporators are below the compressor, run the suction pipe as shown in the sketch (solid lines). But

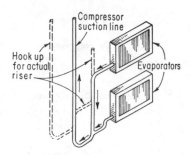

Fig. 9-7. Loop suction line to form a seal for oil with evaporators below compressor.

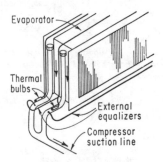

Fig. 9-8. Use separate suction lines when the evaporator has two expansion valves.

if a dual riser is needed, run the suction pipe as shown by the dashed and solid lines.

Q   Where an evaporator has two or more expansion valves, show how to arrange piping so that one valve circuit does not affect the other.

A   See Fig. 9-8. As shown here, run each line several feet before connecting it to the common suction line.

Q   Show how to install piping for a room-cooling unit where the compressor room is hot or where the local code requires a room-ventilation hookup.

A See Fig. 9-9. Here cooling water on the way to the condenser passes through the unit cooler, which is fitted with a fan. This fan runs when the compressor does.

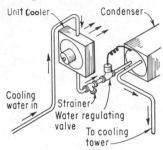

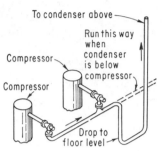

Fig. 9-9. Cool the compressor room with a unit cooler hooked up like this.

Fig. 9-10. Compressors in parallel can use this horizontal discharge line.

Q How should piping be arranged for parallel operation where the compressors are close together and they discharge to one condenser above them?

A See Fig. 9-10. If the discharge line can be run close to the floor, install it as shown in the sketch. If the line cannot be close to the floor, run it as shown in Fig. 9-11. The loop shown is used where the condenser is above the compressor.

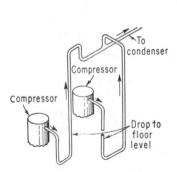

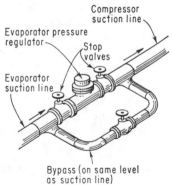

Fig. 9-11. Run loops to floor when compressors in parallel discharge to high main.

Fig. 9-12. Bypass on same level as suction line saves time.

Q What piping hookup would you make to allow work on vital parts of the backpressure regulator without shutting down the system?

A See Fig. 9-12. Run a bypass around the backpressure regulator with a stop valve in the bypass. This bypass should be on the same level as the suction line.

Q   Show the proper hookup for a flooded-coil refrigeration system.

A   See Fig. 9-13. Evaporators are known as the flooded type when liquid refrigerant covers all the heat-transfer surfaces. They are called dry when a portion of the evaporator area is used for superheating the refrigerant. A flooded evaporator uses a float valve, while a dry evaporator uses a thermostatic expansion valve. In the sketch shown, there are a surge drum, evaporator coils, and a liquid float valve. The liquid refrigerant enters through the float-type expansion valve, flows downward from the surge drum, then upward into the evaporator coils. As the liquid flows upward through the coils, heat from outside boils the liquid so that a mixture of liquid and vapor bubbles from the tubes into the surge drum. The function of the surge drum is to separate the liquid from the vapor. The liquid returns to the evaporator, and the vapor returns to the compressor. The surge drum also separates the flash gas which develops in the expansion valve so that this flash gas does not add to the pressure drop in the evaporator. Because the liquid is in contact with all the evaporator surfaces during normal operation, this type of evaporator uses its surface area effectively. Actually, the flooded evaporator is an oil trap. Any oil carried out of the compressor should be removed in the compressor discharge line.

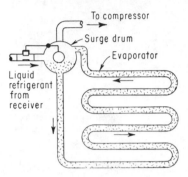

Fig. 9-13. Flooded coil should be hooked up like this.

Q   Show the piping hookup for a shell-and-tube liquid chiller having the refrigerant in the tubes.

A   See Fig. 9-14. A thermostatic expansion valve usually feeds liquid

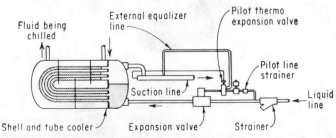

Fig. 9-14. Shell-and-tube direct expansion liquid chiller with piping hookup.

into the tubes as shown. The fluid to be cooled flows through the shell around the baffles. This increases the turbulence and over-all heat-transfer coefficient. This evaporator (chiller) is called the dry type because by using a thermostatic expansion valve some of the tubes superheat the refrigerant vapor rather than evaporate the liquid.

Q How does a shell-and-tube liquid chiller having refrigerant in the shell differ from one having refrigerant in the tubes?
A See Fig. 9-15. A float valve actuates the liquid refrigerant valve to the chiller. The tubes in which the liquid is being chilled are kept cov-

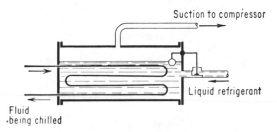

Fig. 9-15. Hookup for liquid chiller with float control.

ered with liquid refrigerant by the float control valve. The suction line to the compressor runs from the top of the chiller and keeps removing the vapors as they boil off in the cooling process.

Q Show how to make a hookup for more than one evaporator.
A Figure 9-16 shows a typical application with two "high-temperature" evaporators and one "low-temperature" unit. A common suction line

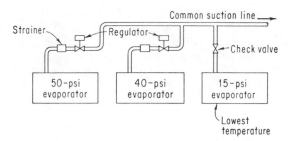

Fig. 9-16. Multiple evaporator installation has regulators on high-temperature units.

serves the three evaporators. The low-temperature unit has a check valve instead of a regulator.

Q   Show the hookup for solenoid valves as used in brine systems to control flow.

A   Figure 9-17 shows two typical applications of solenoid valves for brine service. In the top sketch, valves are used for stops; a thermostat

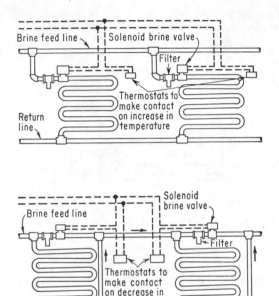

Fig. 9-17. Hookup of two solenoid valve applications.

in the room or compartment being cooled shuts the valve to stop the flow when the desired temperature is reached. Thermostats may be set to maintain different temperatures in different rooms. In the bypass hookup in the lower sketch the flow is shunted past the coil when it reaches the desired temperature. The thermostat controls the valve operation. In both hookups shown, one brine pump serves all coils in the system.

Q   Can a solenoid valve be used for compressor capacity control?

A   Yes, if it is designed for this service. Figure 9-18 shows how the valve is installed in a bypass around one or more of the compressor cylinders. A check valve is used to separate bypassed cylinders from the active ones. A solenoid valve may be controlled by either an automatic or manual switch as indicated.

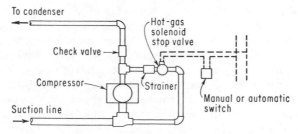

Fig. 9-18. Compressor cylinder bypass is solenoid valve connected.

Q What materials are used for the piping and fittings of refrigeration systems?
A Piping materials for refrigerants, brine, and water include steel, wrought iron, malleable iron, cast iron, copper, and bronze. Piping is made to specifications approved by American Standards Association Code for Pressure Piping, ASA B31.1 (Fig. 9-19). Ferrous pipes are used in

Material	Specification	Allowable stress, internal, psi
Steel, seamless, Grade A	ASTM A53, A120, A106, A83, A179, A192, or API 5L	12,000
Steel, seamless, Grade B	ASTM A53, A106, or A210	15,000
Steel, seamless, alloy, Grades T18, T19	ASTM A213	18,750
Steel, electric-resistance-welded, alloy, Grades T18, T19	ASTM A249	15,900
Steel, top-welded	ASTM A53, A106, A120, or API 5L	9,000
Steel, butt-welded	ASTM A53 or A120	6,800
Steel, electric-resistance-welded	ASTM A135, A178, A214, or A226	10,200
Wrought iron, top-welded	ASTM A72 or API 5L	8,000
Wrought iron, butt-welded	ASTM A72	6,000
Cast iron, pit cast	AWWA	4,000
Cast iron*	FSEC WW-P-421	6,000
Brass, seamless, red brass	ASTM B43	4,700
Copper, seamless	ASTM B42	4,000
Copper tubing, seamless	ASTM B88	4,000

* Centrifugally cast or cast horizontally in greensand molds.

Fig. 9-19. These are approved materials for piping and fittings in refrigeration systems.

ammonia systems; types K and L copper tubing are often used for halocarbons. Valves and fittings for refrigerant piping systems come in a large number of different types, some of which are shown here. Line shutoff valves must give safe, tight closure. Diaphragm packless valves with flare or solder connections are popular for halocarbons. Ammonia valves have extra-long packing glands to prevent leaks; back-seating permits repacking under pressure. Relief valves protect equipment from excessive pressure. Because halocarbon refrigerants are good solvents, they remove sludge, chips, rust, etc., from inner walls of pipes and equipment. For this reason filters are often located in pipelines. Driers are packed with a dehydrating agent to remove moisture from the refrigerant. Economizers use suction gas to subcool liquid refrigerant on its way to the expansion valve. Other important fittings include oil separators, mufflers, liquid-receiver valves, compressor shutoff valves, purgers, etc.

Q   Why is a valve placed in the compressor discharge line?
A   Such a valve isolates the compressor from the rest of the system so that the machine may be shut down and worked on without interference from the refrigerant.

Q   What are safety valves, where are they placed, and what is their purpose?
A   Safety valves are placed where pressures are highest, as at the condenser and receiver. They relieve excessively high pressure in the system, and thus prevent injury to people, equipment, and property should the pressure exceed a safe value. Each valve is set to blow at a predetermined pressure, and the setting adjustment is sealed to prevent tampering.

Q   What is a water-regulating valve and where is it placed?
A   A valve actuated by compressor discharge pressure is placed on the water inlet to regulate the amount of water entering the condenser in accordance with the variation in discharge pressure. When pressure rises, the valve opens to admit more water; conversely, when pressure drops, the valve closes enough to bring the pressure up to normal, thus reducing water flow and thereby lowering operating costs.

Q   Why does a thermostatic expansion valve have a feeler-bulb thermal element strapped to the suction line adjacent to the evaporator?
A   This device ensures passage of superheated vapor to the compressor suction. It prevents liquid refrigerant from getting to the compressor.

Q   What is the suction valve's purpose?
A   It isolates the compressor from the system's low-pressure or suction side. During the pumping down process, this valve is left open until all

refrigerant has been removed from the low side. Then it is closed for safety when working on the low side.

Q Why is a high-pressure–low-pressure cutout necessary?
A It protects the compressor against high discharge and low suction pressures. It consists of an electrical relay that opens the motor circuit. The pressure settings are adjusted in the field so that operation is within the load range.

Q Explain why the refrigerant charging connection is put on the system's high-pressure side.
A Charging on the low side might damage the compressor because slugs of liquid refrigerant could find their way to the compressor suction. The connection is close to the receiver, where refrigerant may be stored. Thus liquid slugs do not reach the compressor.

THERMAL INSULATION

Q Why should bare surfaces be insulated?
A By retarding the flow of heat, thermal insulation will (1) reduce heat gain or loss from piping, equipment, and structures, thus decreasing

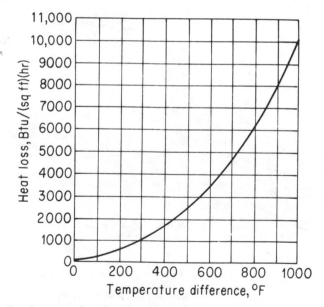

Fig. 9-20. Heat losses in still air from bare surfaces.

the amount of heating or refrigeration capacity needed; (2) control surface temperature for personnel protection and comfort; (3) prevent icing or water vapor condensation on cold surfaces; and (4) make it easier to control space and process temperatures.

Insulation can also (1) add structural strength to floors, walls, or ceilings; (2) prevent or retard the spread of fire; (3) reduce the noise level; and (4) absorb vibration. It is also a good support for surface finish and helps keep out water vapor. The chart in Fig. 9-20 gives heat losses from bare surfaces.

Q What properties must be considered when selecting a low-temperature insulation?

A There are many, such as resistance to (1) heat flow, (2) moisture, (3) decomposition, (4) flame, and (5) vermin. Weight and price are also extremely important. Cold surfaces are insulated to keep heat from entering a system or space operating below ambient temperature. Since heat flows from higher to lower temperatures, it readily moves from warm outside areas into a cold system. Heat gain is more costly to deal with than heat loss. Extracting heat from a refrigerated space may cost ten times more than making up heat loss from a high-temperature system

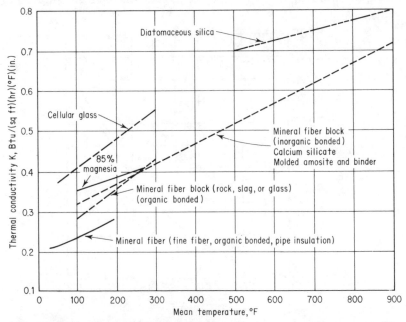

Fig. 9-21. Typical range of K factors for some materials.

would cost. So it generally pays to use much thicker insulation at low temperatures than at high.

Q How does thermal insulation retard the flow of heat?

A Thermal insulating materials are divided into two basic groups: (1) those designed to impede heat conduction (mass type), and (2) those which reduce heat radiation (reflective type).

You can demonstrate heat conduction by holding one end of a metal rod in a fire. Heat gradually travels up the rod to your hand, indicating that dense and heavy material is not generally good thermal insulation. No known material can completely stop the flow of heat as long as there is a temperature difference between two sides. The only perfect nonconductor of heat is a vacuum, but air is almost as good *if* convection currents can be eliminated.

Q Explain thermal conductivity.

A Thermal conductivity (K factor) is the prime measure of a material's ability to transmit or retard heat. The lower the K factor, the better the material's insulating properties. The chart in Fig. 9-21 shows how the K factor varies among some typical insulating materials and with temperature level. For example, thermal conductivity ranges from a low of 0.2 for mineral wool to a high of 0.8 for diatomaceous silica.

Q What are some of the modern insulations used for cold surfaces?

A In recent years plastics have made considerable headway as low-temperature insulation. The largest use to date has been for refrigerated storage rooms and equipment. Other materials used include cellular glass, Fig. 9-22, and the mineral fibers.

Cork and animal or vegetable fibers are now almost entirely out of the picture. Because of the importance of moisture resistance, the emphasis in selecting materials is changing. The data in Fig. 9-23 are important when selecting insulation for cold surfaces.

Q Is moisture a problem in the insulation of cold spaces?

A Yes. Moisture (water vapor) must be kept from moving from the warm to the cold side by means of vapor barriers. Since it is almost impossible to get a perfect vapor barrier, the insulation must also have good moisture resistance. Ideally, it should (1) absorb no moisture or readily give up any that enters and (2) not deteriorate. Thus in low-temperature work the most desirable insulation may not be the one with the finest thermal properties, unless it also happens to have the required moisture resistance. The resistance of an insulation to vapor flow is expressed in terms of its vapor permeability.

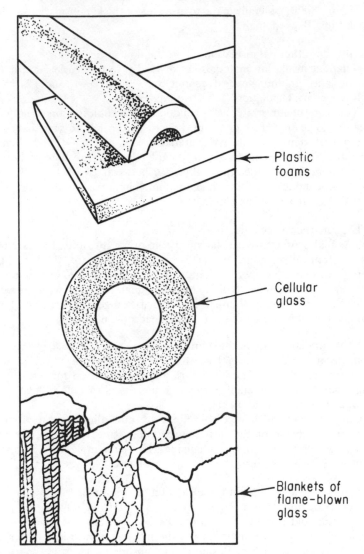

Fig. 9-22. Low-temperature insulation today involves plastic foams, cellular glass, and some mineral fibers.

Temperature range, °F	Material classification	Density, lb per cu ft	Thermal conductivities (K factor), Btu per hr per sq ft per °F per in.
	Plastic foams:		
−275 to 210	Urethane (polyurethane isocyanate)	1.8 to 2.2	0.09 @ 200°F; 0.17 @ 100°F
−200 to 175	Polystyrene	1.0 to 4.0	0.26 @ 40°F; 0.28 @ 75°F
−85 to 120	Urea formaldehyde	0.8 to 1.2	0.20 @ 100°F
−40 to 250	Phenolic	0.33 to 5.0	0.20 @ 70°F; 0.24 @ 250°F
−40 to 160	Polyvinyl chloride	4.5 to 26	0.26 @ 75°F
−40 to 200	Cellular rubber	3.5 to 20	0.24 @ 25°F; 0.30 @ 100°F
−50 to 1600	Expanded silica (perlite)	4.0 to 10	0.33 @ 0°F; 0.38 @ 75°F
−400 to 800	Glass (cellular)	9.0 to 18	0.36 @ 25°F; 0.42 @ 100°F
−250 to 1800	Mineral fibers (rock, glass or slag)	0.5 to 10	0.27 @ 25°F; 0.30 @ 100°F
−300 to 180	Cork	7 to 10	0.26 @ 25°F; 0.29 @ 100°F
−150 to 200	Vegetable & animal fibers	10 to 20	0.26 @ 25°F; 0.30 @ 100°F

Note: Values of temperature, density, and thermal conductivity are approximate. For individual design problems consult insulation manufacturer for design data.

Fig. 9-23. Data used for the selection of thermal insulation.

Q Explain why a vapor barrier is so important?
A Insulation for temperatures below 32°F is only as good as its vapor barrier. A vapor-soaked material has little insulating value. Water vapor is present in air as a gas and will penetrate any material that absorbs air. Thus when an insulated body is at temperatures above ambient (room temperature), the vapor pressure of water is higher at the insulation's inner surface than in the surrounding air. This pressure differential drives vapor outward from the inner surface, Fig. 9-24, through the insulation, and into the outside air.

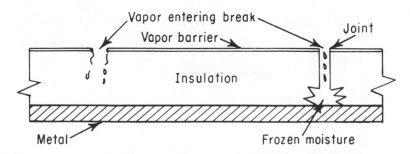

Fig. 9-24. Trouble starts when vapor leaks in through damaged vapor barrier or through improperly sealed joints.

Q   How do reflective surfaces retard heat flow?

A   Highly reflective surfaces such as polished metal prevent heat loss or heat absorption by radiation. Thin aluminum foil is sometimes used as an insulating material in the walls of structures. Shiny metallic roofs also minimize the radiant heating effect of the sun. A thermos bottle contains both a vacuum, which prevents conduction of heat, and highly polished surfaces to prevent heat radiation. Insulation in this general category is extremely light and may be as thin as a single sheet of foil. It also comes in multiple layers and can be attached to other materials.

Surfaces of ordinary building materials have an emissivity and heat-ray absorption greater than 90 per cent. Because of air's low density, conduction through wall space is only 7 per cent. The second sketch in Fig. 9-25 shows the inner surface lined with a reflective-type aluminum foil insulation. Here, radiation is reduced to 7 per cent. The third sketch shows four sheets of emissive metal dividing the wall into three reflective spaces. Heat loss by radiation has dropped to 2.2 per cent and convection to 7.2 per cent, but conduction is up slightly, to 7.2 per cent. However, total heat loss is only 16.6 per cent of the loss before air space was created with reflective foil.

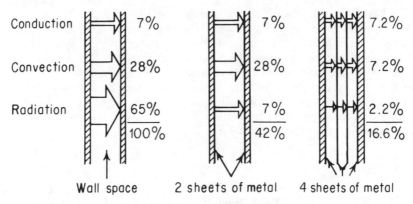

Conduction   7%   7%   7.2%

Convection   28%   28%   7.2%

Radiation   65%   7%   2.2%
              100%   42%   16.6%

Wall space   2 sheets of metal   4 sheets of metal

Fig. 9-25. Heat travels freely through uninsulated wall, while reflective surface reduces radiation.

SUGGESTED READING

Reprints sold by *Power* magazine, costing $1.20 or less
  Piping, 16 pages
  Valves, 24 pages
  Refrigeration, 20 pages
  Power Handbook, 64 pages
  Gaskets, 20 pages
  Conditioned Air for Industry, 16 pages
  Thermal Insulation, 24 pages

Books

Elonka, Steve, and Joseph F. Robinson: *Standard Plant Operator's Questions and Answers*, vols. I and II, McGraw-Hill Book Company, New York, 1959.

Elonka, Steve: *Plant Operators' Manual*, rev. ed., McGraw-Hill Book Company, New York, 1965.

Elonka, Steve, and Alonzo R. Parsons: *Standard Instrumentation Questions and Answers*, vols. I and II, McGraw-Hill Book Company, New York, 1962.

10

LUBRICATION

Q How are moving parts in a refrigeration system lubricated?
A Crankcase oil from the compressor supplies the needed lubrication. In a reciprocating compressor a gear-type oil pump, driven by the crankshaft, lifts oil from the crankcase and sends it through an oil hole in the crankshaft and connecting rods to the crank and crankpin bearings. Oil splashed by the crank action lubricates the cylinder walls and pistons. Compressor valves and all automatic valves in the system receive some crankcase oil carried with the refrigerant. The problems are to lubricate all the compressor parts and to circulate enough oil through the system to supply an oil film on all the working valve parts. This circulated oil must be controlled or the condenser, receiver, and evaporator will become oil-logged.

Q Describe the force-feed lubrication system shown in Fig. 4-3.
A A shaft-driven gear pump supplies oil under pressure to the crankpin and piston-pin bearings and to the shaft seal. A portion of the oil discharged by the pump is bypassed to a filter, from which it returns directly to the crankcase. Oil from the shaft seal and from the filter lubricates the main bearings. A shaft-driven mechanical force-feed lubricator furnishes lube oil to each cylinder wall through tubing piped to the cylinders.

Q Describe a splash lubrication system.
A In splash oiling, lubricating oil is splashed to all bearings by the cranks and connecting rods, which dip into oil carried in the compressor crankcase. Part of the oil splash is thrown into pockets which supply oil to the main bearings, piston-pin bearings, and the shaft seal. The connecting rods of some small machines have dippers and check valves. The speed with which the big ends of the connecting rods dip into the oil forces oil to the crankpin bearings and to the piston-pin bearings through tubes extending the length of the rods. Since proper lubrication of splash-oiled parts depends on the amount of oil splashed by the cranks, rods, and other moving parts, keeping the right oil level is important. If the level is so low that moving parts cannot reach the oil or can dip into

124

it only slightly, there won't be enough oil splashed. If the level is too high, there will be excessive churning of oil, high oil temperature, leakage, and high oil consumption. When that happens, so much oil may reach the cylinder walls of vertical machines that large amounts of lubricant can pass into the high side of the system. Most splash-lubricated machines have sight glasses to check the oil level (Fig. 4-3).

Q How is the amount of oil circulated by a splash lubrication system controlled?
A By controlling the oil level in the crankcase and by having an oil trap on the discharge side of the compressor and keeping the trap drained (Fig. 10-1). The higher the oil level, the greater the amount of oil

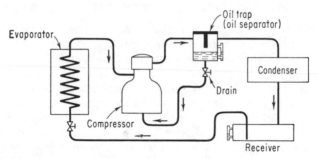

Fig. 10-1. Oil trap (oil separator) on discharge side of compressor can be drained back to the crankcase.

splashed upon the cylinder walls and pistons. This excess oil is carried through the discharge valves where the refrigerant vapor carries it into the system. The discharge valves should be just barely moist with lube oil if the crankcase oil level is at the correct height. One indication of too high a crankcase oil level is a heavy oil coating on the top of the piston when the cylinder head is removed. Oil reaching the evaporator settles to the bottom and must also be drained from time to time. Not all low-pressure systems have oil traps. Oil reaching the evaporator is separated from the suction vapor by a drain and flows back to the crankcase.

Q What problems does oil in the evaporator cause?
A Oil in the evaporator of any type of refrigeration system tends to thicken and coat the inside of the evaporator walls. This reduces the heat transfer through the tubes. The lower the evaporator temperature, the greater the tendency for oil to congeal. Wax often separates out of the oil at very low temperatures. Oil mixing with the refrigerant also varies the boiling temperature, giving the compressor more work to do.

Q   What are the advantages of a force-feed lubrication system?

A   More accurate control of oil distribution is possible. The amount of oil delivered to the high-pressure side is predetermined, thus preventing the variations that occur in the splash system. Another advantage is that less oil foaming takes place on each starting cycle.

Q   What points should be given attention in a force-feed lubricating system for either ammonia or Freon-12?

A   There are six: (1) keep the oil level below the bottom of the connecting rods; (2) keep the compressor free from gas binding; (3) provide oil at higher than crankcase pressure even if bearings are worn; (4) supply lubrication to the stuffing boxes or mechanical seal at pressure no higher than crankcase pressure; (5) add fresh oil without shutting down or admitting air; and (6) strain or filter oil at the pump suction.

Q   Explain how the cylinders are lubricated in refrigeration compressors.

A   Cylinders of most vertical compressors are open to the crankcase. Lube oil that escapes from the main and crankpin bearings is thrown onto the cylinder walls by the centrifugal action of the cranks. Oil is also splashed up from the connecting rods passing through the crankcase oil. Oil landing on the cylinders is carried to the upper portion of the walls by the reciprocating action of the piston rings. The amount of oil supplied in this way to the cylinders of small and medium-sized compressors is more than enough. To keep the oil reaching the upper part of the cylinders to a minimum, oil-control piston rings are used. These rings scrape part of the oil back to the crankcase. In larger compressors the cylinders are farther from the crankshaft. Thus less oil reaches the walls. On large compressors, mechanical force-feed lubricators supply oil directly to the cylinder walls. This supplements the oil from the crankcase. Cylinders of horizontal and rotary compressors receive their entire supply of lube oil from mechanical force-feed lubricators. To keep down the quantity of oil carried out of the compressor and into the system by the discharge gas, oil fed to the cylinders by either method should be held to the minimum needed to protect against wear.

Q   Why is it desirable to use a double-trunk piston-type compressor rather than the crankcase admission type for Freon-12?

A   This is because of the miscibility of Freon-12 with lubricating oil. Here the force-feed lubricator is more practical, as the oil level can be carried much lower than in the splash type. This helps to keep oil out of the system.

Q   What lube oil precautions should be taken before operating a compressor?

A   Make sure that the oil pump is working properly when starting the compressor. Oil gage pressure must be higher than the suction pressure as soon as the compressor gets up to speed. There is an adjustment for regulating the lube oil pressure.

Q What problem will overexpansion cause in the crankcase?

A A common difficulty caused by overexpansion is refrigerant flooding back to the compressor. If the refrigerant liquid reaches the crankcase, it vaporizes or boils out rapidly from the oil and carries oil out past the piston rings with the vapor. This deprives the crankcase of lube oil.

Q What causes oil to foam in the compressor?

A Reduced crankcase pressure at the beginning of each on cycle always causes some foaming; the amount depends on the refrigerant used. Methyl chloride or Freon-12 will foam more than sulfur dioxide or ammonia because more oil is soluble in these first two. As the pressure is lowered, thousands of tiny bubbles (foam) form through entire body of oil in the crankcase. Because this foam has more volume than the solid oil, the level rises, often filling the crankcase.

Q How are compressor bearings oiled?

A Compressor bearings are lubricated by (1) circulation oiling, (2) splash oiling, and (3) ring or chain oiling. In all these systems the same oil is used over and over again. An older method is oiling by mechanical force-feed lubricators or by drip-feed cups. The outboard bearings of some compressors are ring- or chain-oiled. A ring or chain rides on the journal and dips into oil carried in a reservoir in the bearings pedestal. Rotation of the journal causes the ring or chain to rotate and to carry oil up and onto the journal. Grooving and chambers cut into the bearing cause proper distribution of the oil over the full length of the bearing. The oil level must be high enough for the ring to pick it up. If it is too high, oil will leak out at the ends of the bearing.

Q What are some problems in lubricating cylinder walls?

A A vertical refrigeration compressor is subjected to low temperatures at the suction ports and to high temperatures near the cylinder head. Since viscosity decreases with temperature, oil near the suction ports will have much higher viscosity than oil near the cylinder head. The oil must be spread in a thin film over the entire working surface by the piston rings. Oil must help seal the piston rings against the blow-by of refrigerant vapor. To prevent wear, oil must adhere to cylinder walls. The oil should withstand discharge temperatures as high as 250°F.

Q What causes oxidation in lube oil?

A The rate of oxidation depends directly on the quantity of air present and its degree of mixture with oil, on temperature, on the nature of contamination, and on the oil's resistance to oxidation. Little oxidation takes place in vertical compressors having closed crankcases. Deposits on valves and piston crowns are often attributed to lube oil when actually they are caused by impurities returning with suction vapor from the piping system. In large crosshead compressors warm oil does mix with air. Here a chemically stable oil that resists oxidation must be used.

Q   What general characteristics are needed in refrigeration lubricants?
A   See Fig. 10-2. Specific requirements differ with each refrigerant, but some are common to all refrigerants. Most vital is freedom from water. One reason is that water makes most refrigerants corrosive. A second reason is that moisture carried to the expansion valve will freeze, restricting flow. If moisture freezes in the evaporator coils, partial or complete choke will cut down efficiency. The pour point of the oil must be below the lowest temperature to which it may be exposed. Oil must not deposit wax and must separate easily from the refrigerant. A completely distilled straight naphthenic oil with all traces of wax and moisture filtered out is best. For low-temperature service solvent-refined oil of the highest viscosity index that's possible with a low pour point is needed. And last, the oil should not vaporize easily at compression temperatures.

Q   How are petroleum lubricants classified?
A   Petroleum lubricants are classified according to the broad services for which they are widely used: (1) circulating oils, (2) gear oils, (3) machine or engine oils, (4) refrigeration-grade oils, (5) spindle oils, (6) steam-cylinder oils, and (7) wire-rope lubricants. They are also classified according to their properties: (1) greases of calcium, sodium, aluminum, lithium, or barium base; and (2) solid and synthetic lubricants.

Q   How does lubrication reduce friction?
A   Lubrication reduces friction by forcing an oily film between surfaces in order to separate them completely. Then the only friction is from the motion within the oily fluid, which causes the fluid to split into two layers. The top layer sticks to the top surface shaft, and the bottom layer adheres to the lower surface bearing. Each layer travels at a lower speed, shearing the layers on either side of it.

Q   For what are lubricants tested?
A   Lubricating oils are tested for (1) viscosity, (2) flash and fire points, (3) pour point, (4) carbon residue content, (5) emulsification and demulsibility, and (6) dropping or melting point.

Q   What is the viscosity of a lubricating oil?
A   Viscosity is a measure of how thick the oil is—how long it will take a measured quantity of oil to flow through a standard orifice at a given temperature. A viscosity of 120 Saybolt seconds at 100°F means it takes

Type of compressor	ASHRA No.*	Refrigerant	Chemical formula	Atmospheric boiling point, °F	Behavior in contact with mineral lubricating oil	Primary uses	Viscosity, SSU at 100°F†
Reciprocating	717	Ammonia	NH_3	−28	Solubility fairly low. If impure, may form sludge	Ice manufacture	300–320
Reciprocating	744	Carbon dioxide	CO_2	−109.3	Solubility very low. Not reactive	Cold storage, dry ice manufacture	300–320
Reciprocating	764	Sulfur dioxide	SO_2	14.0	Solubility low if oil highly refined. Forms sludge and gum if oil not highly refined	Electric refrigeration	150–160
Reciprocating	40	Methyl chloride	CH_3Cl	−10.8	Soluble. Mixes to lower viscosity and pour point	Electric refrigeration	300–320
Centrifugal	30	Methylene chloride (Carrene No. 1)	CH_2Cl_2	103.6	Soluble. Mixes to lower viscosity and pour point	Air conditioning	300–320
Centrifugal	11	Trichlorofluoromethane (Carrene No. 2, "Freon-11", Genetron 11, Isotron 11)	CCl_3F	74.8	Soluble. Mixes to lower viscosity and pour point	Air conditioning, industrial refrigeration	300–320
Reciprocating, rotary, centrifugal	12	Dichlorodifluoromethane ("Freon-12", Genetron 12, Isotron 12)	CCl_2F_2	−21.6	Soluble. Mixes to lower viscosity and pour point	Air conditioning, electric refrigeration, cold storage, ice manufacture	150–160 300–320 300–320
Reciprocating, rotary	13	Chlorotrifluoromethane ("Freon-13")	$CClF_3$	−114.6	Solubility low. Forms two-layer system	Low-temperature refrigeration	150–160
Reciprocating, centrifugal	22	Chlorodifluoromethane ("Freon-22," Genetron 22, Isotron 22)	$CHClF_2$	−41.4	Soluble at high side temperature. Dependent upon temperature, type oil and per cent oil in mixture, forms two-layer system in low side	Air conditioning, low-temperature refrigeration	150–160 300–320
Centrifugal	113	Trichlorotrifluoroethane ("Freon-113", Genetron 113, Isotron 113)	$C_2Cl_3F_3$	117.6	Soluble. Mixes to lower viscosity and pour point	Air conditioning, commercial & industrial refrigeration	300–320
Reciprocating, rotary, centrifugal	114	Dichlorotetrafluoroethane ("Freon-114", Genetron 114a, Isotron 114)	$C_2Cl_2F_4$	38.4	Soluble at high side temperatures. Dependent upon temperature, type oil and per cent oil in mixture, forms two-layer system in low side	Air conditioning, electric refrigeration, industrial refrigeration	300–320

*American Society of Refrigeration and Air Conditioning Engineers' Standard Designation Number.

†These nominal oil viscosities are generally in agreement with manufacturers' normal recommendations. Under certain conditions and for certain units, a higher or lower viscosity may be specified, so manufacturers' recommendation should always be consulted.

Fig. 10-2. Refrigerants commonly used with refrigeration and air-conditioning units. Using the right lube oil for each service is very important. (From Texaco.)

120 sec for a sample of oil at 100°F to flow through the orifice. If the viscosity were 200, for example, it would show that the oil is almost twice as thick, since it would take 200 sec to flow through the same orifice. Since viscosity changes with temperature, the rate of change is denoted by a viscosity index: the higher the viscosity index, the less effect the temperature change has on the fluidity of the lubricating oil.

Q   What is the pour point of a lube oil?

A   The pour point shows how fluid a lubricating oil can be at very low temperatures; the lower the viscosity of an oil, the lower its pour point. The pour point is important in selecting a refrigeration-grade oil. In small-diameter piping, oil having too high a pour point could congeal during an overnight shutdown in cold weather, causing starved bearings when starting up next morning.

Q   What are the flash and fire points of a lubricating oil?

A   The flash point is the degree Fahrenheit at which vapors given off by heating ignite momentarily from a flame. The fire point is the temperature at which the vapors support combustion. These ratings are important on compressor oil or where the oil is to be used at high temperatures. The flash point of most good lube oils is around 300°F.

Q   What is the cloud point of oil?

A   The cloud point is the temperature at which wax will begin to separate from the oil. Wax separates from the oil as separate crystals, thus thickening the oil. As these crystals form, their opaque white color gives the oil a cloudy appearance. A high-grade refrigeration oil is almost clear. Numerous good oils have a yellowish tinge. Some are almost water white.

Q   How do halocarbon refrigerants affect lube oil?

A   All halocarbon refrigerants thin lube oil greatly. This means that a heavier oil must be used with them. Thinning of oil by halocarbon refrigerants also reduces the pour point and cloud point of the oil. Refrigerants like sulfur dioxide, ammonia, and carbon dioxide do not have a thinning effect on oil.

Q   What characteristics must lube oil have for an ammonia system?

A   An SAE 10 or 20 oil is usually best, the higher grade being preferred for splash-lubricated machines to reduce the risk of foaming. Heavier oil is for pressure lubrication to avoid too much oil being thrown into the cylinders. Heavier oil is also for machines with separate cylinder lubrication. Some large, slow-running machines need SAE 30 for the cylinder and for the bearings, if supplied by drop or other intermittent oiler. Vaporization is not a serious problem because temperature at the end of compression is low in most installations—usually under 300°F or even under 200°F.

Q What characteristics must lube oil have for a carbon dioxide system?
A Viscosity for this system ranges from SAE 20 to SAE 40. These oils are usually one grade heavier than for an ammonia system. Where oil separation is good, standard compressor or engine oils may be used regardless of pour point. But if oil might carry over into the evaporator, a wax- and moisture-free refrigeration oil is needed.

Q What characteristics must lube oil have for a sulfur dioxide system?
A Use a standard SAE 20 refrigeration oil for the glands of open compressors. If the machine is enclosed, use a light oil—between 90 and 200 seconds Saybolt Universal (SSU) at 100°F, depending on the design of the machine. Always take the manufacturer's advice for lubricating these compressors.

Q What characteristics must lube oil have for a methyl chloride system?
A Use an SAE 10 or 20 oil for lubricating enclosed compressors and for glands, cylinders, and bearings of open compressors.

Q What characteristics must lube oil have for the Freon family of refrigerants?
A Freon applies to a family of compounds of carbon, fluorine, chlorine, and hydrogen. Freon refrigerants have a common property of miscibility with oil. In this way their behavior is the same as that of methyl chloride. When used with reciprocating compressors, the same lube oil and precautions apply.

Q What causes lube oil to sludge?
A In compressors where bearings and cylinders are separately lubricated, crankcase oil may sludge from contamination or overheating. If the machine has an enclosed crankcase and the oil is in contact with refrigerant vapor, first check these causes. Contamination is caused by dust, scale, or moisture. Sludge might occur from chemical reaction between oil and some classes of refrigerants or from overheating caused by excessively high compressor temperature. High temperature can be caused by leaking valves, overcharging the system, improper cooling, or oil that is too old. To prevent age causing trouble, follow the manufacturer's recommendations for regular oil changes.

Q What causes deposits in a refrigeration system?
A Deposits can be caused by using oil with too high a pour point or containing excessive wax, overlubrication, or improper use of the oil separator. Deposits are costly as they form an insulating layer inside the piping, which cuts down refrigeration efficiency. Sludge may prevent the regulating valve from working properly. Check for deposits if the compressor has been overheating. Poor-quality oil may also cause sludge.

Q   What conditions may cause deposits in the compressor?
A   The most likely causes are overheating, overlubrication, use of an oil that is too heavy, and corrosion caused by moisture or air in the system. Moisture in carbon dioxide or in sulfur dioxide will also cause corrosion. A chemical analysis is usually needed to find the exact cause of deposits.

Q   What conditions cause excessive oil consumption?
A   The causes are many—some similar to the reasons why an automobile engine wastes oil. Look for worn pistons, rings, and cylinders; abrasive deposits; stuck rings; improper lube oil; too high an oil level in crankcase; too light an oil; or oil carried by refrigerant and dropped out of solution in system. Oil is often lost owing to violent foaming when the backpressure is changed rapidly. Oil absorbs some refrigerant vapor, amount increasing with pressure. If pressure is reduced too suddenly, absorbed vapor is released so quickly that oil can be forced bodily out of the crankcase. In systems having a pressure-equalizing line between the crankcase and suction line, oil may be forced out by high-pressure vapor leaking past the pistons.

Q   What care should be given lubricating oil?
A   Keep oil containers at room temperature for at least 24 hr before opening to preserve the moisture-free qualities of the lubricant. Use small sealed cans if possible so that each may be emptied completely into the compressor. Oil from an open drum that has been resealed is often contaminated with moisture from air entering the drum. Save any oil remaining after topping a compressor by transferring it immediately to a smaller bottle. Make sure that there is no air space before resealing. The best practice is to use such oil for some other purpose and to use only fresh oil for the compressor.

Q   Explain how to add oil to the crankcase of a splash-lubricated vertical compressor.
A   Attach a hose to the filling pipe on the compressor, and dip the free end of the hose in a pail of oil. Shut the refrigerant suction valve while the unit is running, and open the crankcase pump-out valve. That will pump a vacuum in the crankcase. Crack the valve in the filler line and draw in the required amount of oil, being careful to keep the end of the hose immersed in the oil. Close the valve before the pail is empty to avoid admitting air into the system.

Q   Explain how to add oil to the crankcase of a forced-circulation compressor.
A   Attach two hoses to the oil piping, one on each side of the oil pump. (If the machine doesn't have these two connections, use the same method as for a splash-lubricated machine.) Immerse both hoses in a pail of oil.

Close the valve in the oil suction line (between crankcase and pump), and at once open both valves to the two hoses. Oil will be sucked from the pail through one hose and discharged into the pail through the other. As soon as an air-free flow is established, close the valve to the hose on the discharge side of the pump, thereby forcing the oil into the compressor. When the needed amount of oil has been pumped into the unit and before the pail of oil is empty, close the valve to the hose on the suction side of the pump. Then open the valve in the oil suction line from the crankcase.

Q Explain how to drain oil from a vertical compressor that has cylinders open to the crankcase.

A To drain crankcase oil, first close the suction valve, start the compressor, then open the valve in the crankcase pump-out line. This creates a partial vacuum in the crankcase. Run the compressor for several minutes until the vacuum causes the refrigerant to boil out of the oil. Shut down the compressor, and close the discharge valve. Open the bypass valve to equalize the pressure above and below the piston. Open the oil drain valve and drain the oil. Drain the oil about once a year, depending on oil contamination. In ammonia systems check the color of the oil once a week by looking at the gage glass. But don't rely entirely on the appearance of oil in the gage glass. That oil may be clear while the crankcase oil may be contaminated.

Q Explain how to remove water and dirt from lube oil.

A Use an oil separator. Clean it regularly; a dirty separator is almost as bad as none at all.

Q How should the lubrication components be maintained?

A Ring-oiled bearings must always have enough oil in their reservoir (Fig. 4-3) for the ring to dip into oil. This applies to chain-oiled bearings also, as the only major difference in the two designs is that the chain replaces the ring.

Check sight-feed lubricators every hour that the compressor is running. Count the drops fed, using a stop watch to check the feed rate. Refill the lubricator reservoir before the level goes out of sight in the gage glass. Don't skimp on the oil feed rate—you may save a few pennies today and spend hundreds or thousands of dollars for repairs tomorrow.

If your compressors are splash-lubricated, be sure to choose an oil with a viscosity high enough to prevent excessive oil consumption at compression temperature. Oil that is too light gets by the piston rings and mixes with the refrigerant. See that the oil is stable and resistant to oxidation. This keeps carbon and gum formation low, and prevents plugging of oil passages.

CENTRIFUGAL COMPRESSOR LUBRICATION

Q What care must be given to the lubrication of centrifugal systems?
A High-grade turbine oil is recommended for compressor lubrication. If in doubt, consult the builder of your machine. It's usually most convenient to add oil only when the machine is running, being careful not to overcharge the oil system. The oil level falls as the oil is circulated during starting, but under normal running conditions the oil charge increases 5 to 7 per cent in volume as it absorbs refrigerant.

The usual oil level during operation is about half glass. An electric heater in the oil-pump chamber warms the oil charge during a shutdown and prevents excessive absorption of refrigerant. When the machine is shut down, turn the heater on, then turn it off when the cooling water is turned on before the machine is started.

Under usual operating conditions you'll find these suggestions helpful: (1) replace the complete oil charge once a year; (2) at any time when all or some of the oil charge is removed, replace it with fresh oil from a sealed container; (3) after shutdowns of more than one month, remove the bearing covers and add 1 qt of oil to each bearing well before starting the machine; (4) replace the oil filter periodically, depending on the operation and the condition of the filter; (5) at least once a year, inspect and clean the oil strainer.

Q How is a centrifugal compressor system drained of oil?
A To drain the oil system, allow the machine temperature to rise, or artificially heat the machine until the temperature of the cooler is 75°F. This corresponds to atmospheric pressure for Freon-11. Drain the pump and atmospheric float chambers. This removes practically all the oil. The portion of the charge which remains in the bearing wells and seal reservoir is useful for keeping the bearings in good condition and acts as a seal.

Q Explain how faulty lubrication systems contribute to machine failures.
A Out of 150 accidents surveyed by the Hartford Steam Boiler Inspection and Insurance Company, 11 failures were caused by low oil level, 8 by dirty oil or plugged piping, 6 by failure of the lubricator, 5 by an oil leak at the seal or at another point, and 2 by a closed valve in the oil line.

Q What attention must be given to the storage and handling of lubes?
A An improperly protected lubricant supply is almost certain to become contaminated by dust and dirt. No plant man worthy of the name needs to have the dangers of dirt in lubricating oils and greases pointed out to him.

Outdoor storage (Fig. 10-3) is never good. Brand markings and labels wash off or become obliterated. Future dispensing is needlessly endangered and other lubricant contamination becomes possible. Temperature changes from season to season can produce enough expansion and contraction stress over a period of time to cause leaks in the container seams to develop.

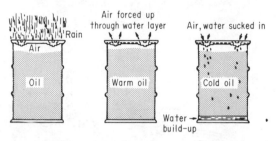

Fig. 10.3. Outdoor storage, never desirable, makes drum contents an easy prey to contamination by outside moisture.

Moisture contamination is a potent threat with outdoor storage. Rain water collects readily inside the chime top (first sketch). It is gradually drawn past the bung by the drum's breathing during alternate hot and cold periods. If you must store lubricants outdoors, spread tarpaulins over the drums or erect a temporary shelter. Turn each drum over so that the bung end is down. If you can't do this, lay the drums on their sides.

Four major recommendations for outdoor storage are: (1) provide at least temporary shelter against the weather; (2) place drums so they cannot breathe through the bungs; (3) make sure the bungs are tight before you move drums; and (4) before opening, dry and wipe the bungs thoroughly, as well as the surrounding surfaces.

Q How should oil be dispensed from storage drums?

A Hand oilcans, large dispensing safety cans, oil wagons, grease guns, lubricating cars, and sump pumps are the most common pieces of dispensing and transporting equipment for grease and oil. Clean this class of equipment regularly. But don't limit your cleaning to mere wiping with solvent-soaked rags. Instead, use generous quantities of solvent for both cleaning and rinsing. Baths of safety solvent (flash point above 100°F) meet the usual safety regulations if ventilation is adequate and correct covers are provided.

The transfer of oil from container (drum) to dispenser may be done with nothing more than a hand-operated drum pump (Fig. 10-4). Such a pump is a positive-delivery type capable of supplying measured oil quantities. You can get pumps with spring-action closable returns, and in air-operated or electrically driven designs. Drum faucets are also efficient and economical transfer devices. They come in different sizes for fast and slow oils.

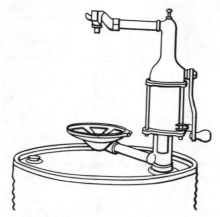

Fig. 10-4. Hand-operated drum pump.

STUFFING BOXES, SHAFT SEALS, AND GASKETS

Q How are refrigerant and oil kept from leaking out of a refrigeration system, and how is air kept from leaking in?
A Jam-type packing, metallic packing, floating packing, and mechanical seals are used to seal between reciprocating and rotary shafts and the crankcase. Gaskets are used to seal flanges and other piping joints.

Q Describe one kind of jam-type packing used successfully in reciprocating ammonia compressors.
A See Fig. 10-5. Soft metallic L-shaped rings fit between waterproof hydraulic flax rings. Flax carries a saturant (lubrication) and provides resiliency. A hollow-center end ring of wrapped fabric and rubber acts as an expansion-compression joint. By pulling up on the gland when the rod is cold, the hollow-center ring stops leakage. When the rod warms, the hole takes up the expansion and reduces friction. A lantern ring is often used with this set for best results.

Q Describe a popular design of mechanical seal used on smaller commercial compressors.
A See Fig. 10-6. This is the metal bellows type of seal. The bellows is attached and sealed to the shaft housing on one end. The seal nose

at the other end bears against a shoulder on the rotating shaft. The thrust is from the spring action of the bellows, backed up by the loading spring. These keep the thrust constant under pressure changes. Then the

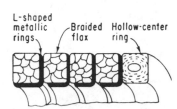

Fig. 10-5. Jam-type packing sets come in many different combinations.

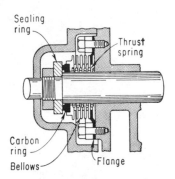

Fig. 10-6. Metal bellows type is common for smaller refrigeration compressors.

power output of the shaft does not vary with pressure changes within the system. The bellows must resist torsional stresses as it prevents the sealing face from revolving with the shaft. Metal bellows are designed so the mean diameter of the seal nose bearing surface equals the mean diameter of the bellows. Then the seal is balanced to pressure so that variation in either internal or external pressures won't affect the sealing thrust exerted by bellows and spring. For example, if a seal is designed to exert 10 psi thrust when installed, it keeps this thrust whether the differential pressure across the seal is 5 psi or 100 psi. There is no horizontal shaft wear because sealing is done by a carbon ring against a metal shoulder, which is vertical to the shaft.

Q   Describe the free-floating metallic packing used on large reciprocating shafts to seal in refrigerant gas.
A   See Fig. 10-7. Most metal packing sets have two segmented rings. Each set is held together by a garter spring stretched around an outside groove. One ring is slit tangentially to the rod; the other is slit radially; and each ring serves a different purpose. Sealing is done by the tangential ring; as pressure is applied, segments are pushed against the case and slide over each other, gripping the rod tighter and tighter. The radially slit ring, known as the cover ring, keeps gases from getting to the seal ring. This is all it does, but without it the metal seal wouldn't work. Rings are made of various metals, depending on the kind of gases they are to seal. Metallic packing rings must be free floating, which means

free to move in their grooves. The number of rings in a set depends on its application—as many as eight are used. These packings need no adjustment. As pressures go higher, packing rings seal tighter and tighter about the rod. They also make up for wear in the same way.

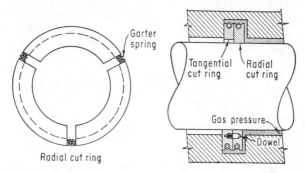

Fig. 10-7. Floating ring packing sets are used on large reciprocating compressors.

Q Describe the combination oil-bellows mechanical seal used in large centrifugal refrigerating compressors.

A See Fig. 10-8. This seal needs no attention. It is built like a mechanical seal, but the wearing faces are held apart while the machine is run-

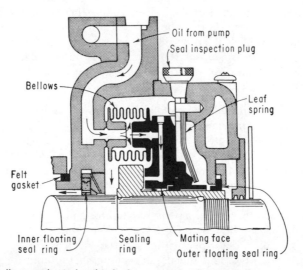

Fig. 10-8. Bellows mechanical seal is for large centrifugal compressor.

ning. The reason there is no wear is that oil pumped between the sealing faces does the sealing, and the oil pump is driven by compressor shaft at the other end of the machine. The faces come together and hold almost a complete vacuum only when the machine is shut down. When the compressor starts, oil is pumped through internal passages in the compressor. Oil flows through the filter and to the inside of the doughnut-shaped seal bellows. Oil pressure expands the bellows, moving the stationary sealing seat against the spring pressure to the stop position that is on the seal housing. This leaves a 0.050-in. opening between the faces of the sealing and the mating ring. Oil in the bellows then passes through drilled passages in the stationary seal to an eccentric oil groove in the face of the stationary sealing seat. Here oil seals the sealing space between the two faces and passes in opposite directions. Oil flows across the outer section of the stationary sealing face and into the space between the shaft and bellows assembly. Flow from this space is restricted by an inner floating ring. The flow continues across the inner section of the sealing face, through the clearance between the rotating sealing faces, and through the clearances between the rotating sealing-seat hub and the stationary sealing seat. This flow is restricted by the small clearance of the floating ring between these two parts. It is also restricted by the atmospheric or outer floating ring between the seal housing cover and the rotating seal hub. The pressure-regulating valve in the oil system remains closed on starting. Thus, full capacity of the oil pump is assured for initial oil supply. During operation the oil passing between the seal faces seals the shaft against inward leakage of air, so metal-to-metal contact is not needed to do the sealing.

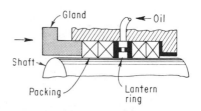

Fig. 10-9. Lantern ring allows lubricant to be fed into the packing set.

Q Are lantern rings used in compressor stuffing boxes between turns of packings in jam-type packing sets?
A See Fig. 10-9. On both old and newer ammonia compressors jam-type packing is often used. An oil lantern ring (spacer) placed in the set allows oil to feed under pressure from the oil pump into the stuffing box. This keeps the shaft and packing lubricated and also helps make a tighter seal so that refrigerant won't leak out.

SUGGESTED READING

Reprints sold by *Power* magazine, costing $1.20 or less
  Bearings and Lubrication, 32 pages
  Gaskets, 24 pages

Mechanical Packing, 24 pages
Mechanical Seals, 24 pages
Piston Rings, 24 pages

Books

Elonka, Steve, and Joseph F. Robinson: *Standard Plant Operator's Questions and Answers,* vols. I and II, McGraw-Hill Book Company, New York, 1959.

Elonka, Steve: *Plant Operators' Manual,* rev. ed., McGraw-Hill Book Company, New York, 1965.

Elonka, Steve, and Alonzo R. Parsons: *Standard Instrumentation Questions and Answers,* vols. I and II, McGraw-Hill Book Company, New York, 1962.

11

DEFROSTING

Q What effect does frost have on cooling coils?

A Frost acts as an insulation; it wastes power and causes needless wear in machinery. For example, let us say a cold-storage room is piped with 1,000 sq ft of 2-in. direct-expansion coils. Assume that it is figured on a heat-transfer coefficient from still air to evaporating surface of 2 Btu/ (sq ft) (hr) (°F). The temperature in the room is 30°F and the refrigerant is 5°F. The refrigerating capacity under these conditions would be $1,000 \times 2 \times 25 \div 12,000 = 4.16$ tons. If the coil is covered with 2 in. of fine, dry frost, the coefficient might be as low as 0.9 Btu per hr. So, the capacity of the coil is reduced about 50 to 60 per cent.

Q Does frost on the evaporator coil always have the same insulating quality?

A No. The insulating effect of frost varies. Pure ice is about 20 per cent as effective as cork used for insulation. Coil frost can contain various amounts of air, which increases the insulating value. One inch of frost can have the insulating value of ½ in. of cork. Usually frost that is slightly below 32°F (freezing) is nearly solid ice. That makes it hard to melt. At very low subzero temperatures frost gets light and fluffy; this forms quickly and also can be melted off quickly. Frost has been known to collect to such thickness that its weight pulled down the evaporator coils, causing extensive damage.

Q What is a simple method of defrosting?

A Frost can be removed from evaporator coils by stopping the compressor until all the frost is melted. Domestic refrigerators are often thus defrosted by turning off the switch and opening the door. Often trays of warm water are placed inside to speed the melting. This method takes long and isn't practical for commercial systems, as there may be spoilage.

Q How would you defrost a simple ammonia coil on a direct-expansion system?

141

A   See Fig. 11-1. To defrost an ammonia coil, reverse the compressor discharge by changing the crossover valves so that hot gas flows back into the suction line. The expansion valve must be closed, and hot gas must

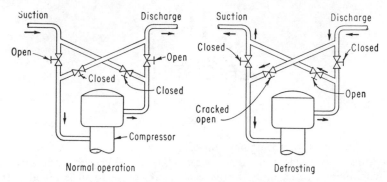

Fig. 11-1. Crossover valves reverse flow through compressor; send hot discharge gas to evaporator.

be fed from the ammonia condensing system by cracking the hot-gas valve. When defrosting, feed the hot gas slowly. A fully opened gas valve may set up heavy expansion stresses where the coils are frozen solid to the pipe supports. While pumping back from the condenser, take care that the cooling water is turned on full or completely shut off and that the condenser is drained, or water may freeze. Make sure that the pressure does not get excessively high in the evaporator coils. After the coils reach 40 to 50°F, it is easy to knock or scrape off the remaining frost or ice formations.

Q   What are some precautions to take when defrosting with ammonia?
A   In ammonia work, the gas line should be ½-in. or ¾-in. extraheavy pipe, as this line may be at condenser pressures. Support the lines well and allow for free expansion through masonry walls or insulation. Install stop valves at the condenser header and keep them closed when the system is shut down. When defrosting, feed hot gas slowly. A fully opened gas valve may set up heavy expansion stresses where coils are frozen solid to the pipe supports. Don't melt off frost and ice formation completely. After the coils reach 40 to 50°F, it is easy to scrape the soft ice off. But be careful; welded joints and corroded pipe have often broken when handled roughly.

BEWARE: Never apply heat to a coil that contains refrigerant if it is closed off from the system, because the refrigerant cannot escape. Many disastrous accidents happen when a steam hose is turned on the coils while both the inlet and outlet of the coils are closed.

Q Explain how to defrost an ammonia or a Freon-12 system with hot gas.

A See Fig. 11-2. Defrosting evaporator coils with the hot-gas method permits defrosting the coils without raising the temperature of the com-

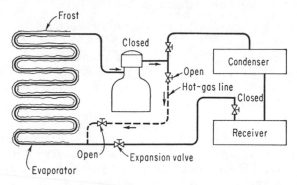

Fig. 11-2. Hot-gas line on new system sends discharge gas to evaporator instead of to the condenser.

partment above 32°F. For this method a hot-gas line is connected from the discharge side of the compressor to a point just beyond the expansion valve in the evaporator coil. Stop valves at either end of the hot-gas line are provided for control. To operate: (1) close the liquid line stop valve ahead of the expansion valve; (2) close the compressor discharge valve; (3) open the valves in the hot-gas line, and start the compressor.

> CAUTION: Open the valves very slowly and not too wide. There is always danger of liquid from the evaporator slugging the compressor badly and causing damage.

Q Some cold-storage plants have warm brine lines running to various coils for defrosting. Explain how to defrost such a system.

A See Fig. 11-3. Use a portable defrosting unit with a 50-gal closed tank mounted

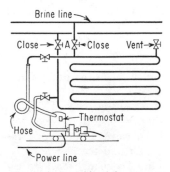

Fig. 11-3. Portable defroster carries hot brine directly to evaporator of large cold-storage system.

on a hand truck. The unit should be hooked to a small, low-head pump, directly connected to a 1-hp motor. Two electric immersion heaters can be fitted into the tank. The hose must be connected from the defrosting unit to defrosting valves on the coil. To defrost, close stop valves A or the coil, isolating coil from system. Circulate brine in the tank (warmed by electric heating elements) through the coil. Isolated coils are easy to defrost by this method. Installing brine connections on coils and electrical outlets is often simpler and less costly than running long warmbrine lines from the heat exchanger to the plant.

Q   Describe a semiautomatic-type defrost device used on commercial systems.

A   One kind of semiautomatic device has a cam for exerting extra pressure on the control spring. This raises the operating temperature above freezing. After defrosting is complete, it must be turned back to the normal operating position.

Q   Describe a full automatic-type defrost device.

A   The full automatic type has a special spring for automatically pulling the control back to the regular operating position as soon as the evaporator temperature is high enough to melt the frost. Some makers install time clocks to defrost the evaporator at 24 hr intervals. Modern defrosting systems are designed to keep temperatures of the storage spaces low enough while defrosting so food won't spoil.

Q   Name three common methods of defrosting evaporator coils.

A   Three methods are (1) hot discharge gas, (2) warm air, and (3) water.

Q   How do the warm air and water methods work?

A   When the coil is defrosted with warm air, it is isolated temporarily from the space which it cools and warm ambient (at room temperature) air is blown over the coil. The water defrost system floods the outside of the coil with water until the frost is melted.

Q   Explain the hot-gas method of defrosting a system having two evaporators.

A   See Fig. 11-4. This method sends gas from the compressor discharge directly into the evaporator. To defrost the evaporator A, close valve 1, open valve 3, and open valve 2. Evaporator A now operates as a condenser and receives hot gas, condensing it and releasing it to the liquid line. To change back to cooling operation, close valve 3, close valve 2, and open valve 1.

CAUTION: Open valve 1 slowly to prevent liquid from surging back to the compressor.

In the two-evaporator system shown the system receives heat from one evaporator and also from work of compression during defrosting. In a single-evaporator system heat from hot-gas defrost must come from work of compression alone, unless energy from the discharge gas is stored during refrigeration operation and used to assist the defrost.

Defrosting can be a time-consuming job in buildings and institutions where many small units are scattered about in various departments.

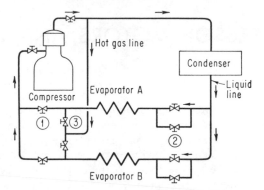

Fig. 11-4. Hot-gas method of defrosting a system having two evaporators.

Q   Are there other ways of defrosting besides those named here?
A   Yes. Electrical resistance heaters are placed against the evaporator coil to heat the coil after passing low voltage current through the metal. Another method is to spray warm water over the frost after stopping the unit.

SUGGESTED READING

Reprints sold by *Power* magazine, costing $1.20 or less
    Refrigeration, 20 pages
    Air Conditioning, 20 pages
    Power Handbook, 64 pages

Books
    Elonka, Steve, and Joseph F. Robinson: *Standard Plant Operator's Questions and Answers*, vols. I and II, McGraw-Hill Book Company, New York, 1959.
    Elonka, Steve: *Plant Operators' Manual*, rev. ed., McGraw-Hill Book Company, New York, 1965.
    Elonka Steve, and Alonzo R. Parsons: *Standard Instrumentation Questions and Answers*, vols. I and II, McGraw-Hill Book Company, New York, 1962.

12

COMPRESSOR MOTORS AND DRIVES

Q   What are the three major requirements in selecting a source of power to operate the compressor in a refrigeration system?

A   The first requirement of a compressor drive should be reliability. The refrigeration system usually protects goods whose value is many times the cost of the system itself, so the system must be dependable. The second requirement is economy, both in initial cost and in operating cost. This includes maintenance and repair costs, as they will be a recurring item during the life of the system. The third requirement for a compressor drive is simplicity. This means the drive should be simple to operate, simple to maintain, and simple to repair. If the equipment is easily understood, it is usually better operated.

Q   What features have made electric motors so widely used in refrigeration systems?

A   The three points of reliability, reasonable economy, and simplicity of operation have led to widespread adoption of electric drives for refrigeration compressors. Reliability of operation has been designed and built into motors of all sizes and type. Power systems have been integrated so that the power shutdowns of the past occur very infrequently and only for very short periods. Electric motors will operate for long periods with a minimum of attention if kept clean and dry. Space requirements are usually small for the horsepower required, and various types of motors are available to meet the requirements of different systems.

Q   What are the two basic types of a-c electric motors used for compressor drives?

A   Electric drives for compressors are divided into single-phase and polyphase motors. Nearly all compressors of 1 hp or under (fractional horsepower) are single phase, though single-phase motors are available in ratings up to 5 hp. For special applications, polyphase motors are also available in fractional horsepower sizes, and nearly all units of 5 hp and over are polyphase.

Q   What is the main difference between single-phase and polyphase motors?

A   See Fig. 12-1. A single-phase motor has only one winding in which the alternating current flows to produce a magnetic field that alternates

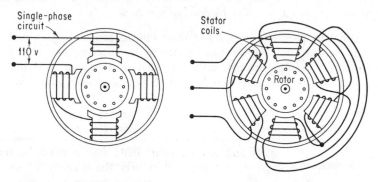

Fig. 12-1. Schematic diagram of single-phase motor windings.

Fig. 12-2. Schematic diagram of three-phase motor windings.

in polarity but does not rotate. That is, it has no starting torque and tends to act like a single-cylinder engine at dead center. If a slight auxiliary torque is applied, the motor will start in either direction and soon accelerate up to normal speed. Nearly all single-phase induction motors have some auxiliary means of starting. Polyphase motors have two or more windings (one for each phase) distributed evenly around the circumference of the stator. The alternating current flowing through these coils produces a magnetic field that rotates as the alternating current changes in strength and direction. This rotating magnetic field, by magnetic attraction, pulls the rotor around with it, developing torque.

Q   What is a squirrel-cage induction motor?

A   See Fig. 12-2. The squirrel-cage motor has two main parts: the stator, sometimes called the field, contained in the stationary frame; the rotor, sometimes incorrectly called the armature, mounted on the motor shaft and free to rotate. These terms (rotor, stator) are more accurate when referring to a-c equipment, since some large units have a revolving field and a stationary armature. The stator of a polyphase motor has a separate winding for each phase imbedded in the iron laminations that make up the stator, with each winding connected to one wire of the supply circuit. The rotor of the squirrel-cage motor consists of iron laminations pressed on to the motor shaft and slotted to receive the rotor windings or squirrel-cage conductors. All rotor windings are electrically and mechanically connected by end rings. A rotating magnetic field is developed in the stator laminations. This has a definite polarity, depending on the momentary direction of current flow (see Fig. 12-2). It creates an induced current in the rotor with a similar polarity. As the field rotates in the stator, the rotor poles follow this field at a speed approaching synchronous speed. Some slip is required; that is, the rotor must lag behind the rotating field by some percentage in order to develop power.

Q   What methods are used for starting squirrel-cage induction motors?
A   All squirrel-cage motors can be started by applying line voltage to the motor windings through either manual or magnetic switches. In larger sizes starting currents may be quite high and some type of reduced-

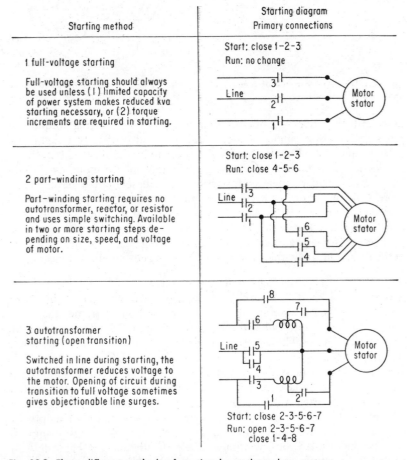

Starting method	Starting diagram Primary connections
1 full-voltage starting Full-voltage starting should always be used unless (1) limited capacity of power system makes reduced kva starting necessary, or (2) torque increments are required in starting.	Start: close 1-2-3 Run: no change
2 part-winding starting Part-winding starting requires no autotransformer, reactor, or resistor and uses simple switching. Available in two or more starting steps depending on size, speed, and voltage of motor.	Start: close 1-2-3 Run: close 4-5-6
3 autotransformer starting (open transition) Switched in line during starting, the autotransformer reduces voltage to the motor. Opening of circuit during transition to full voltage sometimes gives objectionable line surges.	Start: close 2-3-5-6-7 Run: open 2-3-5-6-7 close 1-4-8

Fig. 12-3. Three different methods of starting large three-phase motors.

voltage starting equipment may be required by the public utility (see Fig. 12-3).

Q   What is a slip-ring or wound-rotor motor, and where is it used?

A   The basic difference between squirrel-cage and wound-rotor motor is in the construction of the rotor. Both designs use the same stator. The resistance in a rotor winding has a direct effect on the operating characteristics of the motor. High starting torque and low starting current require a high-resistance rotor winding, while low slip at full load requires for high operating efficiency a low resistance which can be controlled to meet operating conditions. By increasing rotor resistance, the motor will develop maximum starting torque for minimum starting current. As the rotor speed increases, resistance is reduced until the rotor windings are shorted out and the motor operates as a squirrel-cage induction motor. By varying the rotor resistance the speed of the motor can be controlled down to about 50 per cent of its full speed. Wound-rotor motors are often used where speed variation is needed.

Q   What are the four types of single-phase motors from the standpoint of starting methods?

A   See Fig. 12-4. The four main types of single-phase motors are (*a*)

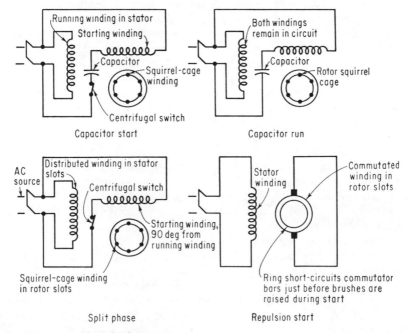

Fig. 12-4. Types of single-phase motors.

capacitor start, (*b*) capacitor run, (*c*) split phase, and (*d*) repulsion start. Once up to full speed they all operate as induction motors. In a capacitor-start motor the stator contains an auxiliary winding and centrifugal switch in series with the capacitor. During starting the capacitor holds the auxiliary winding enough out of phase to create a low starting torque. When the rotor reaches about 75 per cent of its speed, the centrifugal switch opens the auxiliary circuit.

In the capacitor-run motor the capacitor and auxiliary winding always remain in the circuit. This makes it a type of two-phase motor, which is quieter and vibrates less than the split phase.

Split-phase motors are the simplest design of single-phase motors but have a low starting torque and a high starting current. They are usually applied to small fans and similar objects. Split-phase motors have a squirrel-cage rotor and two stator windings spaced 90 electrical degrees apart. Auxiliary winding poles are halfway between the main winding poles, and the winding is so designed that the current in it is out of phase with the main winding. This differential is enough to create a magnetic field similar to a two-phase motor's. At about 75 per cent of full speed the centrifugal switch opens the circuit to the starting or auxiliary winding.

The repulsion-start motor, sometimes called a repulsion-induction motor since it operates as an induction motor at full speed, has wound-rotor coils whose terminals are brought out to a commutator. The commutating brushes are placed so that the induced magnetic field in the rotor coils has the same polarity as that in the stator field coils. It is therefore repelled by the stator field, causing the rotor to turn. The commutator bars are shorted out at about 75 per cent of full speed, and the motor operates as an induction motor. Repulsion motors are used where high starting torques are required.

Q   What is a hermetically sealed motor?
A   A hermetically sealed motor, or hermetic unit (see Chap. 4, "Compressors"), when applied to compressor drives can include any one of the above-mentioned types of induction motors completely sealed within a casing. No make-or-break contacts can be placed within the casing because the arcing of contacts could cause oil or refrigerant breakdown. Starting relays that are operated by the starting surge are used. On starting, the surge of current is enough to close the magnetic contactor that connects the starting winding to the circuit. As the motor speed increases, the current decreases to a point where the contactor opens and removes the starting coil from the circuit. This relay is mounted outside the sealed housing. The leads are brought out through insulated pressure-sealed bushings.

Q What is a synchronous motor?

A See Fig. 12-5. A synchronous motor is one that rotates at the same speed as the alternator supplying its current or at a definite fixed relation-

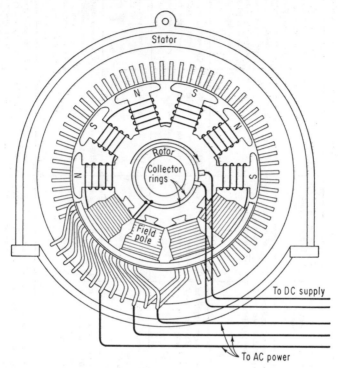

Fig. 12-5. Engine-type synchronous motor.

ship to that speed. The speed is determined by the frequency of the supply circuit and the number of poles in the rotor, based on the formula that motors are built in the range from 2 poles (3,600 rpm at 60 cycles) to 100 poles (72 rpm at 60 cycles). Construction differs from induction motors mainly in that the rotor has a number of insulated coils wound on pole pieces. These coils are connected to a separate source of direct current, usually supplied by a motor-generator set (exciter), that creates alternate north-south poles on the rotor when the direct current is applied. The pole faces also contain damper or amortisseur windings similar to squirrel-cage windings on an induction motor. The synchronous motor is started by first applying alternating current to the stator windings. This immediately creates a revolving magnetic field as in an induction motor. This revolving field reacts with the squirrel-cage wind-

Polyphase Motors

Type	Speed regulation	Speed control	Starting torque	Breakdown torque	Applications
General-purpose squirrel-cage (Design B)........	Drops about 3% for large to 5% for small sizes	None, except multispeed types, designed for 2 to 4 fixed speeds	100% for large, 275% for 1-hp 4-pole unit	200% of full load	Constant-speed service where starting torque is not excessive. Fans, blowers, rotary compressors, and centrifugal pumps
High-torque squirrel-cage (Design C)...........	Drops about 3% for large to 6% for small sizes	None, except multispeed types, designed for 2 to 4 fixed speeds	250% of full load for high-speed to 200% for low-speed designs	200% of full load	Constant-speed where fairly high starting torque is required infrequently with starting current about 550% of full load. Reciprocating pumps and compressors, crushers, etc.
High-slip squirrel-cage (Design D)...........	Drops about 10 to 15% from no load to full load	None, except multispeed types, designed for 2 to 4 fixed speeds	225 to 300% of full load, depending on speed with rotor resistance	200%. Will usually not stall until loaded to max. torque, which occurs at standstill	Constant-speed and high starting torque, if starting is not too frequent, and for high-peak loads with or without flywheels. Punch presses, shears, elevators, etc.
Low-torque squirrel-cage (Design F)...........	Drops about 3% for large to 5% for small sizes	None, except multispeed types, designed for 2 to 4 fixed speeds	50% of full load for high-speed to 90% for low-speed designs	135 to 170% of full load	Constant-speed service where starting duty is light. Fans, blowers, centrifugal pumps, and similar loads
Wound-rotor............	With rotor rings short-circuited, drops about 3% for large to 5% for small sizes	Speed can be reduced to 50% by rotor resistance. Speed varies inversely as load	Up to 300% depending on external resistance in rotor circuit and how distributed	300% when rotor slip rings are short circuited	Where high starting torque with low starting current or where limited speed control is required. Fans, centrifugal and plunger pumps, compressors, conveyors, hoists, cranes, etc.
Synchronous............	Constant	None, except special motors designed for 2 fixed speeds	40% for slow to 160% for medium-speed 80% pf. Specials develop higher	Unity-pf motors 170%; 80%-pf motors 225%. Specials up to 300%	For constant-speed service, direct connection to slow-speed machines and where power-factor correction is required

D-C and Single-phase Motors

Type	Speed regulation	Speed control	Starting torque	Breakdown torque	Applications
Series.........	Varies inversely as load. Races on light loads and full voltage	Zero to maximum depending on control and load	High. Varies as square of voltage. Limited by commutation, heating, and line capacity	High. Limited by commutation, heating, and line capacity	Where high starting torque is required and speed can be regulated. Traction, bridges, hoists, gates, car dumpers, car retarders
Shunt.........	Drops 3 to 5% from no load to full load	Any desired range depending on design, type of system	Good. With constant field, varies directly as voltage applied to armature	High. Limited by commutation, heating, and line capacity	Where constant or adjustable speed is required and starting conditions are not severe. Fans, blowers, centrifugal pumps, conveyors, wood and metal-working machines, elevators
Compound.........	Drops 7 to 20% from no load to full load, depending on amount of compounding	Any desired range, depending on design, type of control	Higher than for shunt, depending on amount of compounding	High. Limited by commutation, heating, and line capacity	Where high starting torque and fairly constant speed is required. Plunger pumps, punch presses, shears, bending rolls, geared elevators, conveyors, hoists
Split-phase.........	Drops about 10% from no load to full load	None	75% for large to 175% for small sizes	150% for large to 200% for small sizes	Constant-speed service where starting is easy. Small fans, centrifugal pumps, and light-running machines, where polyphase is not available
Capacitor.........	Drops about 5% for large to 10% for small sizes	None	150 to 350% of full load depending on design and size	150% for large to 200% for small sizes	Constant-speed service for any starting duty and quiet operation, where polyphase current cannot be used
Commutator-type.........	Drops about 5% for large to 10% for small sizes	Repulsion-induction, none. Brush-shifting types, 4 to 1 at full load	250% for large to 350% for small sizes	150% for large to 250% for small sizes	Constant-speed service for any starting duty where speed control is required and polyphase current cannot be used

Fig. 12-6. Motor characteristics and applications.

153

ings in the rotor, causing the rotor to rotate and come up to nearly synchronous speed. At this point, the direct current is applied to the rotor coils, creating alternate north-south poles that lock in step with the revolving magnetic field of the stator.

Q   What do the following terms mean: (1) Starting or break-away torque, (2) accelerating torque, (3) pull-up torque, (4) pull-in torque, (5) synchronous torque, and (6) pull-out torque?
A   (1) Starting or break-away torque is that developed by the motor at the instant of starting; (2) accelerating torque is that developed between standstill and pull-in; (3) pull-up torque is the minimum developed between start-up and pull-in; (4) pull-in torque is that developed between slip speed and synchronous speed; (5) synchronous torque is that existing during operation; (6) pull-out torque is that developed when the motor pulls out of step because of overload. Unless especially designed, synchronous motors have low starting torque.

Q   What are some of the advantages of synchronous motors used as compressor drives?
A   Synchronous motors give efficient operation at speeds from 3,600 down to 60 rpm. They can be mounted directly on the shaft of slow-speed loads, eliminating belts, speed-reducing devices, and floor space. Their efficiency at low speed is higher than that of induction motors, and their speed is constant regardless of load variations. Besides driving their mechanical loads, synchronous motors can be used to raise the power factor in a plant. This can be a very important feature if the public utility imposes a low power factor penalty.

Q   What types of enclosures are available for large motors, and where should they be used?
A   Types of motor enclosures include: (1) Dripproof, which is least expensive for clean, indoor installations. Defined by National Electrical Manufacturers Association (NEMA) as a motor having vent openings so constructed that particles falling on machine at any angle not greater than 15° from the vertical cannot enter the machine directly or by running on horizontal surfaces. (2) Splashproof; for more protection, designed so liquid drops or solid particles falling on the unit, coming toward it on a straight line or at any angle not less than 100° from the vertical cannot enter directly or by running along horizontal surfaces. (3) Weather-protected, which has side openings through which air travels up into the space above the stator. The rotor fan draws air from this chamber through the motor windings. Heated air is discharged through stator vents. This type is used for outdoor operation. (4) Pipe-ventilated, which is totally enclosed except for inlet and outlet duct openings. Connecting

ducts admit and discharge ventilating air, which is circulated by a fan within the motor enclosure or by a separate external blower. This type is used where clean air is not available at motor location. (5) Totally enclosed fan-cooled unit, which has a fan at the end of the motor shaft to circulate air around the totally enclosed frame. This type is used for dusty atmospheres or outdoor installations. Lower maintenance and greater reliability offset the initial cost. (6) Explosion-proof, to withstand internal explosion of a specified gas or vapor and to prevent ignition of any gas or vapor surrounding the motor. For detailed data, refer to National Electrical Code, article 500.

Q How is the direction of rotation of a motor changed?

A To reverse the rotation of a three-phase motor, interchange any two of the three leads. In a two-phase motor the connections to one phase should be changed. Single-phase repulsion-start motors are reversed by shifting the brushes according to markings on the motor. In other types of single phase motors the internal connections of the starting circuit may be changed. Manufacturers' advice should be followed.

Q How are large motors connected to the shaft of the compressor or other driven machine?

A Several methods are used for connecting the motor and loads. Where the motor speed is greater than the driven machine speed, flat belts, V belts, chain drives, or gear reducers are used. Where the driven machine speed is greater than the motor speed, as in centrifugal compressors, a geared speed increaser is often used. In the medium range of speeds, 600 to 1,800 rpm, direct-connected induction motors are used. Where synchronous motors are used, the wide range of speeds of these motors makes it possible to mount the motor's rotor directly on the extended crankshaft of the compressor or driven machine. Following is a summary of motor characteristics and applications (see Fig. 12-6).

Q What are the advantages of steam drives, and where are they found?

A Before the advent of modern electric drives and power grids, steam furnished nearly all the power for refrigeration systems. In areas of low-cost fuel, or where steam is used in the process, steam-engine-driven compressors are still used. The steam engine acts as a giant reducing valve, taking steam at boiler pressure and sending the reduced exhaust into the process system to give up more of its heat. This is an efficient way to use steam. Breweries, dairies, laundries, and chemical plants are examples. Steam turbines are finding wide use in large-tonnage air-conditioning systems for the same reason. Public utilities generating large amounts of steam for sale for winter heating loads are able to distribute the same capacity of steam at reasonable rates to operate steam turbine drives for air-conditioning loads in summer. This helps them level ou. their winter and summer load. The output capacity of steam-driven compressors is easily varied to meet the load by changing the compressor speed.

Q   Are internal combustion engines ever used as prime movers in refrigeration systems?
A   Both diesel and gas engines are still installed for operation in areas where this type of fuel is available at low cost. These drives also give direct capacity through speed variation within operating limits of the engine.

Q   What are some of the operating procedures that should be followed to ensure the best operation from V-belt drives?
A   The grooves of the sheaves should be checked frequently for rough spots and uneven wear. Belts should never be removed or installed by rolling or forcing them over the sheave. Always loosen the tension so that belts may be removed or installed easily. Check belt tension periodically, especially when new belts have been installed. Too much tension shortens belt life and puts needless strain on the bearings. Too little tension will allow slippage, causing burning and other damage. Shafts and sheaves should be checked for parallel position and proper alignment. When replacing multiple belt sets, matched sets should always be used. Oil and other lubricants should be kept off belts.

MOTOR CONTROLS

Q   What is the purpose of motor controls for refrigeration and air-conditioning equipment?
A   There are five basic purposes: (1) to admit electrical energy to the motor at a proper rate, (2) to protect against any possible fault in the electrical system which could cause a sudden inrush of current, (3) to prevent overheating of the motor while it is operating, (4) to regulate the motor speed, and (5) to withdraw electrical energy when the need ceases.

Q   What standards and codes are there for motor controls?
A   The National Electric Code, Article 430: Motors, Motor Circuits, and Controllers, covers basic minimum provisions and rules for the use of motor controls. These provisions are a guide for safeguarding persons and buildings and their contents from hazards arising from the use of electricity, but this code is not a design manual. The Underwriters Laboratories, Inc., provide standards for Industrial Control Equipment, No. 508, and for Temperature Indicating and Regulating Equipment, No. 873.

You must also follow the local, city, and state codes, as well as the regulations of the local power company.

NOTE: To understand the controls for any specific machinery, the reader should study the instruction books, circuit diagrams, and literature of the installed equipment.

Q What checks must be made on electric drives?
A Inspect the starter contacts for burning, and replace or reface them if necessary. Check the starter operation and look carefully for loose connections.

Q What attention should be given to the large motors of centrifugal compressors at scheduled maintenance time?
A Clean the motor of foreign material. If bearings are oil-lubricated, drain the old oil and refill. Check end-play for indication of thrust-bearing wear. On variable-speed motors, inspect the drum controller for smooth operation. Be sure to check the resistance elements for any loose connections. Inspect the compressor motor assembly on hermetically sealed machines. Also check the thrust bearing and shaft-journal bearings for wear and proper clearance. Examine the motor winding insulation.

Q What attention do safety controls need?
A Check the operation and settings of all safety controls. That means the condenser high-pressure cutout, the low-refrigerant temperature cutout, the purge high-pressure cutout, the low-chilled-water temperature cutout, and the low-oil-pressure switch. Also inspect the operating controls, such as the chilled-water controller.

Inspect and clean all thermostats, hydrostats, and relays. Check for proper calibration. Examine the sequence of operation of control instruments and operators, such as damper motors and chilled-water valves.

SUGGESTED READING

Reprints sold by *Power* magazine, costing $1.20 or less
    Electric Motors, 32 pages
    Measuring Electricity, 24 pages
    Electrical Protection, 24 pages
    Electrical Practice, 32 pages
    Bearings and Lubrication, 32 pages
    Belts and Chains, 20 pages
    Balancing Machinery, 24 pages
    Vibration Isolation, 16 pages

Books
    Moore, Arthur H., and Steve Elonka: *Electrical Systems and Equipment for Industry*, Van Nostrand Reinhold Company, New York, 1971.

Elonka, Steve: *Plant Operators' Manual*, rev. ed., McGraw-Hill Book Company, New York, 1965.

Elonka, Steve, and Julian L. Bernstein: *Standard Electronics Questions and Answers*, vols. I and II, McGraw-Hill Book Company, New York, 1964.

Emerick, Robert Henderson: *Troubleshooters' Handbook for Mechanical Systems*, McGraw-Hill Book Company, New York, 1969.

13

PRACTICAL CALCULATIONS

Q What is meant by horsepower equivalent? How would you find it?
A Horsepower equivalent in heat is 2545 Btu. It is found as follows:

1 hp = 33,000 ft-lb per min
1 hp-hr = 33,000 × 60 = 1,980,000 ft-lb
1 Btu = 778 ft-lb, so 1,980,000 ÷ 778 = 2545 Btu

Q What is meant by the coefficient of performance?
A This is an index of the performance of a refrigeration machine. It is equal to the ratio of the heat removed to the work needed to remove the heat (or compression). The following formula is used:

$$K = \frac{H_2 - H_1}{H_3 - H_2}$$

K = coefficient
H_1 = total heat of refrigerant before it enters the expansion valve
H_2 = total heat of refrigerant before it enters the compressor
H_3 = total heat of refrigerant as it leaves the compressor

Q Find the coefficient of performance for standard-ton operating conditions using ammonia as the refrigerant.
A Total heat of ammonia at the expansion valve or

$H_1 = 129$ Btu
$H_2 = 616$ Btu
$H_3 = 721$ Btu
$K = \dfrac{616 - 129}{721 - 616} = 4.63$ (ratio of refrigerating effect to the work of compression)
1 hp-hr = 2545 Btu per hr
1 refrigeration ton = 200 Btu per min or 12,000 Btu per hr
Therefore $\dfrac{2545}{12,000} \times 4.63(K) = 0.972$ ton produced per horsepower expended.

Q What is meant by ice-making capacity?

A Refrigeration machines and plants are sometimes rated with respect to their ice-making capacity. The ice-making capacity of a machine is the number of tons of ice which it can produce in one 24-hr day; in general, it is equal to from 50 to 70 per cent of the refrigeration capacity. Heat required to produce ice is that which is necessary to cool the water to the freezing point, to freeze it, to cool it further to the temperature of the brine bath, and to cover other losses.

Q Give an example of the term "ice-making capacity."

A Let us assume the temperature of water is 90°F and the brine temperature is 15°F. Heat which must be removed in cooling the water from 90 to 32°F is equal to the temperature range multiplied by the specific heat of water. The specific heat of water is 1 Btu and the temperature range is $90 - 32 = 58$°F. Heat for cooling the water is then equal to $58 \times 1 = 58$ Btu. To freeze 1 lb of water at 32°F requires the removal of 144 Btu.

The specific heat of ice is about 0.5, so the amount required to cool the ice to the temperature of the brine is equal to the temperature range multiplied by 0.5.

The temperature range is $32 - 15 = 17$°F, so the heat required is equal to $17 \times 0.5 = 8.5$ Btu. The total heat required is $58 + 144 + 8.5 = 210.5$ Btu.

Usually about 15 per cent additional heat is allowed to cover such losses as heat transferred through brine tank insulation, heat from cans, meltage, etc. This additional heat is equal to $210.5 \times 0.15 = 31.5$ Btu.

The total amount of refrigeration required, then, is $210.5 + 31.5 = 242$ Btu per lb of ice; or $242 \times 2,000 = 484,000$ Btu per ton. Thus 484,000 Btu is expended to produce a ton of ice that has a refrigerating effect of only $144 \times 2,000 = 288,000$ Btu. Since a ton of refrigeration is the removal of 288,000 Btu per day, the ice-making capacity is equal to $288,000 \div 484,000 = 0.596 = 59.6$ per cent of the refrigeration capacity.

Q Give the formula used to calculate the refrigeration needed to cool commodities.

A $H = WS(t_2 - t_1)$

Let H = number of Btu of refrigeration effect required

W = weight of stored goods

S = specific heat of stored goods

t_1 = temperature of goods when put in storage

t_2 = temperature of cold-storage room

Q Find the refrigeration needed to cool 20,000 lb of lean beef from a temperature of 95 to 35°F in 24 hr.

Commodity	Per cent water	Specific heat above freezing, Btu/(lb)(°F)	Specific heat below freezing, Btu/(lb)(°F)	Latent heat	Storage-temperature range, °F
Apples............	84	0.90	0.49	122	35–40
Apricots...........	85	...	...	...	35–40
Bananas..........	75	0.90	...	...	55–56
Oranges..........	...	0.90	0.47	124	40–45
Vegetables (mxd)...	89 av.	0.90	0.45	130	40–45
Celery...........	94	0.91	0.46	136	32–34
Carrots..........	86	0.86	0.45	126	35–38
Bacon...........	20	0.50	0.30	28	30–32
Beef (lean)........	72	0.77	0.41	102	30–35
Lamb............	67	0.67	0.30	95	28–30
Pork (fresh).......	46	0.68	0.38	86	30–32
Poultry..........	60	0.80	0.42	85	15–20
Fish (frozen).......	74	0.75	0.40	101	5–15
Butter...........	...	0.64	0.34	15	15

Fig. 13-1. Properties of perishable commodities are needed to calculate refrigeration needs.

A From the table in Fig. 13-1 the specific heat of beef above the freezing point is found to be 0.77, and the difference in temperature between 95 and 35 is 60°F. Substituting these values in the formula $H = WS(t_2 - t_1) = 20,000 \times 0.77 \times 60 = 924,000$ Btu. When 924,000 is divided by 288,000 (Btu in a ton of refrigeration), the quotient is found to be 924,000/288,000 = 3.21 tons.

Q How does time influence refrigeration requirements?
A In the above question, one day (24 hr) is allowed for a product to be cooled. If cooling were to be done in 1 hr, the total Btu would be divided by 12,000 (Btu per hr).

Q How much refrigeration is needed to cool 1 lb of beef, to freeze it, and to cool it further to 5°F? The initial temperature of the beef is 70°F, and the specific heat above the freezing point of 28°F is 0.77 Btu per lb. The latent heat of fusion is 102 Btu per lb, and the specific heat after freezing is 0.41 Btu per lb.
A
Cooling from 70 to 28°F .... $1 \times (70 - 28) \times 0.77 = $ 32.34 Btu
Freezing .... $1 \times 102 = 102.00$ Btu
Cooling to 5°F .... $1 \times (28 - 5) \times 0.41 = $ 9.43 Btu
 Total cooling per pound   143.77 Btu

Q Explain how refrigeration is lost through cold-storage walls.
A Since the temperature of a cold-storage room is several degrees

below that of the atmosphere, heat flows from the outside to the inside through the wall and its insulation. This transfer of heat is continuous. Therefore the rate of flow, or heat transmission, depends on the kind and thickness of insulation, thus constituting a transfer coefficient factor (see Fig. 13-2a).

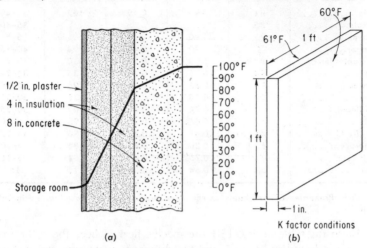

Fig. 13-2. Wall construction has a direct effect on heat-transfer conditions.

Q What is meant by the term "K factor"?

A The K factor is the number of Btu transmitted through 1 sq ft surface 1 in. thick in 1 hr for 1°F temperature difference between the two sides of the surface (see Fig. 13-2b).

Q What is the equation usually used to find the heat loss through storage walls?

A
$$U = \cfrac{1}{\dfrac{x_1}{K_1} + \dfrac{x_2}{K_2} + \dfrac{x_3}{K_3} + \cdots}$$

U = heat in Btu flow per hr through wall per sq ft wall area per °F difference in temperature

$x_1 + x_2 + x_3 \cdots$ = thicknesses of different wall materials, in.

$K_1 + K_2 + K_3 \cdots$ = transmission coefficients given in Fig. 13-3, column 3

Q A wall is made up of 2 in. of yellow pine lined with two layers of 2-in. corkboard and ½ in. of facing cement. Wall is 20 ft high and 40 ft long.

Material	Density, lb per cu ft	Average Btu passing per hr through a plate of material 1 sq ft in area 1 in. thick per °F difference in temperature of the two faces
Seaweed including air..............	...	0.1788
Kapok (loose packed)..............	0.874	0.238
Wool felt........................	20.6	0.363
Cork (ground)....................	4.867	0.3225
Corkboard.......................	4.74	0.305
Celotex 7/16 in. thick.............	13.5	0.33
Sawdust.........................	11.86	0.406
Pine............................	...	1.045
Oak (7/64 in. thick)..............	37.45	2.74
Mineral wool....................	12.50	0.2614
Asbestos paper..................	31.2	1.249
85%-Magnesia...................	13.48	0.5104
Sil-O-cel........................	28.1	0.5805
Facing cement...................	...	1.010
Masonry (average)...............	128.000	4.10
Fire-felt........................	26.000	0.596

Fig. 13-3. Heat loss calculations depend on these data.

Maximum outside temperature $= 98°F$
Inside temperature $= 38°F$
How much heat passes through the wall each hour?

A Coefficient:
Yellow pine $= 1.045$
Corkboard $= 0.305$
Facing cement $= 1.01$

Thickness of material:
Yellow pine $= 2$ in.
Corkboard $= 4$ in.
Facing cement $= \frac{1}{2}$ in.

$$U = \cfrac{1}{\cfrac{2}{1.045} + \cfrac{4}{0.305} + \cfrac{\frac{1}{2}}{1.01}}$$

$$= \frac{1}{1.914 + 13.11 + 0.494}$$

$$= \frac{1}{15.514}$$

$$= 0.0644 \text{ for 1 sq ft and } 1°F \text{ temp difference}$$

Wall area = 20 × 40 = 800 sq ft
Temperature difference = 98 − 38 = 60°F

$H = 0.0644 \times 800 \times 60 = 3091.2$ Btu per hr

To make provision for wall losses and air infiltration, H should be increased 50 per cent for actual conditions:

$H = 1.5 \times 3091.2 = 4636.8$ Btu per hr

The above formula and computations can be used for calculating heat loads for cold-storage refrigeration or for air-conditioning problems.

Q What heat is transmitted through the walls of a room 20 ft long, 15 ft wide, and 10 ft high, with the outside temperature 70°F, the inside temperature 5°F, and the heat-transmission coefficient of the wall equal to 0.10 Btu per sq ft per degree temperature difference per hour?
A Area

2 walls	20 × 10 = 2 × 20 × 10 =	400 sq ft
2 walls	15 × 10 = 2 × 15 × 10 =	300 sq ft
Floor	15 × 10 = 1 × 15 × 10 =	150 sq ft
Ceiling	15 × 10 = 1 × 15 × 10 =	150 sq ft
Total area		1,000 sq ft

Temperature difference (70 − 5) = 65°F
Heat-transmission coefficient = 0.10 Btu
Total heat transfer 1000 × 65 × 0.10 = 6500 Btu per hr
Add 50 per cent for losses 1.5 × 6500 = 9250 Btu per hr

Q Explain some of the heat generated in cold-storage rooms.
A Refrigeration is required to offset the heat generated in the refrigerated rooms by workmen, lights, motors, fans, etc. The average workman in a cold-storage room gives off about 500 Btu per hr to the surrounding air. For each watt of capacity in electric lights, 3.41 Btu should be allowed. The heat of motors, fans, etc., is in proportion to the mechanical equivalent of heat.

Q What is the formula for making an accurate estimate of the number of gallons of water per minute required to be pumped through the condenser?
A $G = \dfrac{(H - g)W}{(T_1 - T_2)8.33}$
G = gpm of water
H = total heat in 1 lb of refrigerant gas entering condenser
g = heat of the liquid leaving condenser
W = lb of refrigerant circulated per min

$T_1 =$ temperature of entering water
$T_2 =$ temperature of leaving water

Q Find the refrigerating effect per pound of R-12 circulated under standard-ton conditions and the required amount of condenser water. Assume an evaporating temperature of 5°F and a superheat of 9°F.

A Total heat in 1 lb R-12 at 5°F = 80 Btu
Total heat in 1 lb R-12 at 86°F liquid − 9°F subcooling = 86 − 9 = 77°F = heat content at 77°F = 25.56 Btu
Refrigerating effect = 80 − 25.56 = 54.44 Btu per lb
Refrigerating effect per lb = 54.44 Btu
1 ton rate = 200 Btu per min
$$\frac{200}{54.44} = 3.67 \text{ lb/(ton)(min)}$$

Water temperature in = 76°F
Water temperature out = 81°F
3.67 lb of R-12 is circulated per min per ton
Substitute in the formula:
$$G = \frac{(80°F - 25.56)\,3.67}{(76°F - 81°F)\,8.33} = 4.8 \text{ gpm per ton}$$

Q What are the desired pressure for condenser water and the maximum circulating rate per ton of refrigeration per minute for Freon-12?
A The desired pressure for condenser water ranges from 30 to 50 psi, with a maximum circulating rate of 3 to 8 gpm per ton.

Q How much heat must be removed from a storage room 100 ft long, 20 ft wide, and 10 ft high to cool the air from 80 to 35°F? The specific heat of air is 0.242. Referring to Fig. 13-4, the weight of air at 35°F is 0.08.

Temperature, °F	Weight, lb per cu ft (saturated)
30	0.081
35	0.080
40	0.079
45	0.078
50	0.077
60	0.076
70	0.074
80	0.0725
90	0.071
100	0.069

Fig. 13-4. Temperature of air (at sea level) has effect on weight.

A $100 \times 20 \times 10$ ft = 20,000 cu ft of air; 1 cu ft weighs 0.080 lb at 35°F; the weight of the air in the room is $20,000 \times 0.08 = 1,600$ lb. $H = W \times$ specific heat $\times T_d = 1,600 \times 0.242 \times (80 - 35) = 17,424$ Btu.

Q   How many pounds of calcium chloride is needed in a brine system? Assume that a brine tank holds 100 cu ft. Suppose that a calcium chloride brine supply is to have a density of 80°F. From Fig. 13-5 it is seen that brine must contain 18.86 per cent of calcium chloride and have a specific gravity of 1.158.

A $100 \times 62.5 = 6,250$ lb
   $6,250 \times 1.158 = 7,237.5$ lb
   $7,237.5 \times 0.1886 = 1,465$ lb calcium chloride

or $\dfrac{1,465}{100} = 14.65$ lb per cu ft

$\dfrac{14.65}{7.5} = 1.95$ lb per gal solution

Salinometer reading at 64°F	Percentage of calcium chloride by weight	Freezing point, °F	Specific gravity at 64°F
28	6.60	25.8	1.050
34	8.49	23.6	1.065
40	9.43	22.1	1.073
48	11.32	19.1	1.089
52	12.26	17.7	1.097
60	14.15	13.8	1.114
64	15.09	11.9	1.122
68	16.03	10.0	1.131
76	17.92	5.4	1.149
80	18.86	3.1	1.158
84	19.80	−0.8	1.167
88	20.75	−4.4	1.176
92	21.69	−8.0	1.186
96	22.63	−11.6	1.196
100	23.58	−15.2	1.205
104	24.52	−19.6	1.215
108	25.46	−24.4	1.225
112	26.40	−29.3	1.236

Fig. 13-5. Data for brine calculations needed by refrigeration engineers.

Q   Explain the comparison of the ice-making capacity of a plant with its tons of refrigeration capacity. Assume that water is available at 80°F. The average temperature of the blocks of ice coming from the evaporator is 22°F. The summertime loss, due to radiation, is 25 per cent.

A   Heat removed to cool 1 lb of water to 32°F = $(80 - 32) \times 1 = 48$ Btu.

Latent heat of fusion of water = 144 Btu
Heat removed to cool ice to 22°F with the specific heat of ice 0.5 =
     (32 − 22) × 0.5 = 5 Btu
Total heat removed from each pound of ice = 197 Btu
Heat removed to make up for radiation is 25 per cent of 197 or
     $49\frac{1}{4}$ Btu
Total refrigeration per pound = $246\frac{1}{4}$ Btu

For 1 ton of ice made under these conditions, 2,000 × $246\frac{1}{4}$ =
492,500 Btu must be removed. Since 1 ton of refrigeration is 288,000 Btu,
under these conditions it requires 492,500/288,000 = 1.71 tons of refrigeration to produce 1 ton of ice.

Q   Under winter conditions the above plant has water available at
36°F, and radiation losses are reduced to 8 per cent. How will ice-making
and refrigeration tonnage compare?
A   Heat removed to cool 1 lb of water to 32°F = (36 − 22) × 1 =
     14 Btu

Latent heat of fusion of water	144 Btu
Specific heat of ice at 0.5 = (32 − 22) × 0.5	5
Total heat removed to produce each pound	149
Heat removed to make up for radiation = 0.08 × 149	11.92
Total refrigeration per pound of ice	160.92

1 ton of ice requires 2,000 × 160.92 = 321,840 Btu
Comparing this with 1 ton of refrigeration at 288,000 Btu, the relationship is

$$\frac{321,840}{288,000} = 1.117 \text{ tons of refrigeration to make 1 ton of ice}$$

Thus only 1.117 standard tons of refrigeration are required per ton
of ice when 36°F water is available, as compared with 1.71 tons per ton
of ice in the previous question.

Q   If your brine tanks had a slight ammonia leak, how would you determine the amount of ammonia in the brine liquor?
A   One way to determine the amount of ammonia in brine liquor is by
titrating a weighed sample of the brine solution against a tenth-normal
sulfuric acid solution, using methyl orange as the indicator for the end
point. The amount present is calculated as follows: Let A equal milliliters
of acid used in titration and B equal the weight of the sample taken; then

$$\text{Per cent ammonia} = \frac{A \times 0.017 \times 100}{\text{sample weight, gal}}$$

For the result in ounces per gallon, multiply the per cent of ammonia
by 1.33 and by the density of the brine solution.

Standard tenth-normal sulfuric acid solution can be purchased from any laboratory supply house. With a little practice the end-point transition of methyl orange from yellow (alkaline) to a deep orange (acid) can be easily detected.

Q   How is the displacement of a refrigeration compressor calculated?

A   The theoretical displacement of a compressor is found by the formula:

$$D = \frac{0.7854 \times B^2 \times L \times N \times \text{rpm}}{1{,}728}$$

where D = theoretical displacement or swept volume
  B = diameter of cylinder, in.
  L = length of stroke, in.
  N = number of cylinders
  rpm = revolutions per minute
  1,728 = cu in. per cu ft

EXAMPLE: Calculate the capacity of a two-cylinder 6 × 6 compressor at 400 rpm.

$$D = \frac{0.7854 \times 6^2 \times 6 \times 2 \times 400}{1{,}728} = 78.5 \text{ cfm}$$

NOTE: Tables have been made up to eliminate most of these calculations, but the methods should be known for use when tables are not available.

Q   What is the displacement of a two-cylinder compressor with $1\frac{7}{8}$-in. bore × $2\frac{1}{4}$-in. stroke operating at 560 rpm?

Bore	Cfm	Bore	Cfm
1	0.0454	1⅞	0.160
1⅛	0.0575	2	0.182
1¼	0.0710	2¼	0.230
1⅜	0.0857	2½	0.2841
1½	0.1022	2¾	0.344
1⅝	0.1196	3	0.409
1¾	0.1390		

Fig. 13-6. Data for determining cylinder displacement.

A   See Fig. 13-6.

$$D = \frac{F \times L \times N \times \text{rpm}}{100}$$

where F = volume factor from the table

$$D = \frac{0.160 \times 2.25 \times 2 \times 560}{100} = 4.032 \text{ cfm}$$

Bore × stroke, in.		Cu in. per rev.	Cfm per 100 rpm	Displacement, cfm, at various rpm										
				200	220	240	260	280	300	330	360	400	450	500
4	× 4	100.5	5,818	11.64	12.80	13.96	15.13	16.29	17.45	19.20	20.94	23.27	26.18	29.09
4½	× 4½	143.1	8,283	16.57	18.22	19.88	21.54	23.19	24.85	27.34	29.82	33.13	37.28	
5	× 5	196.4	11.36	22.73	25.00	27.27	29.54	31.82	34.09	37.50	40.91	45.45	51.13	
5½	× 5½	261.3	15.12	30.25	33.27	36.30	39.32	42.35	45.37	49.91	54.44	60.50	68.06	
6	× 6	339.3	19.64	39.27	43.20	47.12	51.05	54.98	58.90	64.79	70.68	78.54	88.36	
6¼	× 6¼	383.5	22.19	44.39	48.82	53.26	57.70	62.14	66.58	73.24	79.90	88.77	99.87	
6½	× 6½	431.4	24.96	49.93	54.92	59.91	64.91	69.90	74.89	82.38	89.87	99.86	112.3	
7	× 7	538.8	31.18	62.36	68.60	74.83	81.07	87.31	93.54	102.9	112.2	124.7	140.3	
7¼	× 7¼	598.6	34.64	69.28	76.21	83.14	90.07	97.00	103.9	114.3	127.4	138.6		
7½	× 7½	662.7	38.35	76.70	84.37	92.04	99.71	107.4	115.0	126.6	138.1	153.4		
8	× 8	804.2	46.54	93.08	102.4	111.7	121.0	130.3	139.6	153.6	167.6	186.2		
8¼	× 8¼	882.0	51.04	102.1	112.3	122.5	132.7	142.9	153.1	168.4	183.8	204.2		
8½	× 8½	964.7	55.83	111.7	122.8	134.0	145.1	156.3	167.5	184.2	201.0	223.3		
9	× 9	1145	66.27	132.5	145.8	159.0	172.3	185.6	198.8	218.7	238.6			
9¼	× 9¼	1243	71.94	143.9	158.3	172.7	187.1	201.4	215.8	237.4	259.0			
9½	× 9½	1347	77.94	155.9	171.5	187.0	202.6	218.2	233.8	257.2	280.6			
10	× 10	1571	90.90	181.8	200.0	218.2	236.3	254.5	272.7	300.0	327.2			
10½	× 10½	1818	105.2	210.5	231.5	252.6	273.6	294.7	315.7	347.3	378.8			
11	× 10	1900	110.0	220.0	242.0	264.0	286.0	316.0	330.0	363.0	392.0			
12	× 12	2714	157.0	314.0	345.5	376.8	408.4	439.8	471.2					

Fig. 13-7. Data for ammonia compressor calculations. (From ACRMA *Equipment Standards*.)

Pressure and corresponding saturated temperature of suction gas		Discharge pressure, psig															
		135		145		155		165		175		185		195		205	
		Corresponding saturated temperature, °F															
		78.7		82.5		86.2		89.7		93.0		96.2		99.3		102.3	
Pressure, psig	Temperature, °F	Cfm/Ton	Bhp/Ton	Cfm/Ton	Bhp/Ton	Cfm/Ton	Bhp/Ton	Cfm/Ton	Bhp/Ton	Cfm/Ton	Bhp/Ton	Cfm/Ton	Bhp/Ton	Cfm/Ton	Bhp/Ton	Cfm/Ton	Bhp/Ton
0	−28.0	12.34	2.51														
5	−17.2	8.42	2.02	8.64	2.12	8.89	2.22	9.16	2.32	9.40	2.41	9.66	2.50				
10	−8.4	6.36	1.694	6.53	1.782	6.70	1.869	6.84	1.955	7.01	2.04	7.16	2.12	7.35	2.20	7.53	2.28
15	−1.0	5.09	1.449	5.21	1.531	5.34	1.611	5.45	1.690	5.56	1.765	5.67	1.839	5.79	1.911	5.93	1.980
20	5.5	4.28	1.270	4.39	1.341	4.46	1.412	4.55	1.482	4.64	1.552	4.72	1.621	4.80	1.688	4.90	1.755
25	11.3	3.69	1.127	3.75	1.193	3.84	1.259	3.91	1.324	3.98	1.387	4.06	1.449	4.12	1.509	4.20	1.568
30	16.6	3.24	1.011	3.31	1.073	3.38	1.133	3.44	1.193	3.50	1.251	3.56	1.307	3.63	1.363	3.69	1.419
35	21.4	2.92	0.910	2.96	0.971	3.00	1.029	3.06	1.085	3.11	1.140	3.16	1.193	3.22	1.244	3.27	1.293
40	25.8	2.62	0.825	2.66	0.883	2.71	0.938	2.75	0.992	2.81	1.044	2.85	1.094	2.90	1.144	2.94	1.193
45	30.0	2.39	0.755	2.43	0.807	2.47	0.859	2.50	0.910	2.54	0.961	2.58	1.099	2.62	1.057	2.65	1.103
50	33.8	2.18	0.689	2.21	0.739	2.25	0.789	2.28	0.838	2.33	0.886	2.36	0.934	2.39	0.980	2.42	1.026

Note: Based on two-cylinder compressors with 6-in. bore and 6-in. stroke.

Values above heavy line are for ratios of compression above 8.0, and for these conditions it is recommended that multistage compression be considered.

$\frac{\text{Cfm}}{\text{Ton}}$ = average values of swept volume, in cubic feet per minute, required per ton refrigeration based on a suction superheat of 10°F and on liquid ammonia entering the refrigerant control valve at the saturated temperature corresponding to the discharge pressure.

$\frac{\text{Bhp}}{\text{Ton}}$ = average values of actual compressor brake horsepow r per ton refrigeration.

Fig. 13-8. Capacity requirements for ammonia compressors.

Equipment manufacturers usually furnish tables of the capacities of their compressors. If these are not available, the above formulas may be used to calculate average capacities.

Q   What is the capacity of a 9 × 9 two-cylinder compressor operating at 300 rpm with 30 psig suction and 165 psig discharge?

A   See Figs. 13-7 and 13-8. Figure 13-7 shows that a 9 × 9 compressor operating at 300 rpm has displacement of 198.8 cfm. Figure 13-8 shows that at 30 psig suction and 165 psig discharge, 3.44 cfm per ton is required. Therefore 198.8 ÷ 3.44 = 57.7 tons capacity.

Q   What would horsepower requirements for this compressor be?

A   See Fig. 13-8. Under the bhp per ton column, 1.193 bhp per ton is required. 57.7 × 1.193 = 68.5 bhp is required, so a 75-hp motor would be used.

CALCULATING LOST COOLING EFFECT

Q   Following 1 lb of refrigerant through the system, we learn that not all of it lands in the evaporator as liquid. Excess heat in the liquid causes vapor flashing at the expansion valve. How much refrigerant has to go through the cooling coils to keep the load at the right temperature, say 20°F?

A   Property tables show that it takes 553 Btu to vaporize 1 lb of ammonia. Thus, if 5,530 Btu must be removed from a cold room to hold it at 20°F, we have to get 5,530/553 = 10 lb of ammonia liquid into the cooling coils per minute, hour, or day, depending on how fast the heat is to be removed.

Note that we have to feed 10 lb of ammonia liquid into the cooling coil to remove 5,530 Btu from the cold room. Now that isn't the same as saying that we can remove the 5,530 Btu by circulating 10 lb of ammonia through the system. The reason is that not all the liquid passing through the expansion valve in Fig. 13-9 remains liquid. Some of it vaporizes on entering the cooling coils.

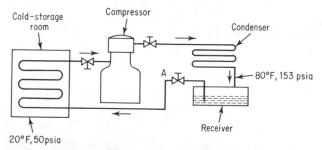

Fig. 13-9. Flash vapor produced when liquid valve A does no useful refrigeration in coils.

Because the receiver shown contains ammonia at 80°F, its pressure (from ammonia tables) is 153 psi absolute. (We must use absolute values when referring to property tables.) With the cooling coil (evaporator) of Fig. 13-9, if the space is to be held at 20°F, property tables show that to maintain this temperature, the compressor must run fast enough to hold the cooling-coil pressure at about 50 psia.

We know that the liquid in the receiver is at 80°F and 153 psia. Thus we must choke back its flow into the cooling coil by throttling the expansion valve. Otherwise the compressor couldn't hold the cooling-coil pressure at 50 psia. Liquid entering the evaporator becomes unstable because it contains too much heat to remain a liquid at the lower pressure of 153 psia. But the compressor pulls vapor out of the cooling coil as fast as it forms.

Q   What about flash vapor in the above problem?

A   From Fig. 13-10 (skeleton from Mollier diagram for ammonia) or from property tables, we find that liquid ammonia at 80°F and 153 psia contains 132 Btu per lb (horizontal scale below point A), and at 50 psia and 20°F it contains 65 Btu per lb (same scale). Subtracting 65 from 132 leaves 67 Btu excess heat in each pound of ammonia going through the throttling valve into the cooling coil.

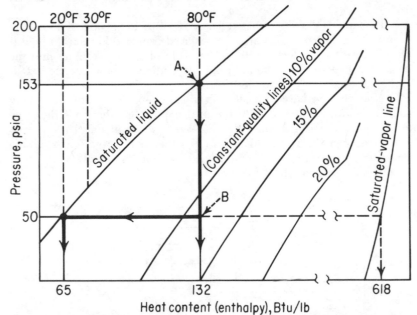

Fig. 13-10. When throttled from 153 psia at 80°F to 50 psia at 20°F by expansion valve, the 67 Btu excess heat in 1 lb causes 12 per cent of liquid to flash into vapor.

This 67 Btu excess is the heat that goes to vaporize part of the ammonia liquid as it enters the lower pressure of the cooling coils.

Property tables show that at 50 psia, 1 lb of ammonia absorbs 553 Btu in changing from liquid to vapor. So if the only heat available is the excess heat, 67 Btu per lb, the amount vaporized is 67/553 = 12 per cent.

Q What is the expansion loss in the above problem?

A In Fig. 13-10, point A represents conditions in the receiver. Follow the heavy line to point B, which represents the new pressure and temperature. Point B is between the 10 and 15 per cent curves (constant-quality lines), or about 12 per cent. This vapor isn't a very good refrigerant because it does not absorb much heat per degree change in temperature. Thus as far as refrigeration value is concerned, we lose 12 per cent of every pound of liquid at the expansion valve.

We have already figured it would take 10 lb of refrigerant to remove the 5,530 Btu. Now it is clear that the 10 lb of refrigerant must actually be in the cooling coil as a liquid. And to get 10 lb of ammonia that has full refrigerating effect, we must circulate 10 × 112 per cent = almost 12 lb. The 100 per cent equals the 10 lb we need, and the 12 per cent equals the vapor flash loss.

Q Can we figure the above problem in another way?

A Yes. Let's start with a pound of refrigerant in the receiver. Upon passing through the throttling valve, 12 per cent of it immediately flashes to vapor. Then 100 − 12 = 88 per cent, so 0.88 lb of the liquid actually settles in the evaporator to do useful refrigerating work. And this 0.88 lb absorbs only 0.88 × 553 = 486 Btu in vaporizing. So, if we start with a total of 10 lb in the receiver, heat absorbed in the evaporator is 10 × 486 = 4,860 Btu. Subtracting 4,860 from 5,530 leaves 670 Btu in the refrigerated space. Therefore, we could not possibly maintain 20°F.

Remember that the flash vapor not only has to be allowed for in figuring the amount of refrigerant needed, but also affects compressor capacity. Since all vapor must be compressed and condensed, the compressor must be large enough to handle both flash vapor and vapor from evaporation.

SUGGESTED READING

Reprints sold by *Power* magazine, costing $1.20 or less
    Power Handbook, 64 pages
    Air Conditioning, 20 pages
    Refrigeration, 20 pages
    Handbook on Fans, 16 pages
    Conditioned Air for Industry, 16 pages

Books

Elonka, Steve, and Joseph F. Robinson: *Standard Plant Operator's Questions and Answers*, vols. I and II, McGraw-Hill Book Company, New York, 1959.

Elonka, Steve: *Plant Operators' Manual*, rev. ed., McGraw-Hill Book Company, New York, 1965.

ASHRAE Data Book, American Society of Heating, Refrigerating and Air-Conditioning Engineers.

14

OPERATION

Q Sketch a typical ammonia system, showing where thermometers are needed for efficient operation.

A Thermometers, or at least thermometer wells, are needed (Fig. 14-1) at (1) refrigerant line leaving condenser, (2) liquid line ahead of expansion valve, (3) space or substance to be cooled, (4) evaporator outlet, (5) all important points where branch suction lines connect into the main suction header, (6) main suction line, and (7) discharge line from compressor.

Q In Fig. 14-1, what does thermometer 1, which shows the temperature of the refrigerant leaving the condenser, tell the operator?

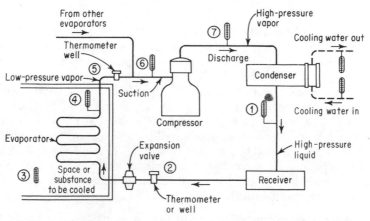

Fig. 14-1. Thermometers should be placed in all the places indicated for efficient operation.

A It indicates if the condenser needs cleaning, if the circulating pump is running as it should, if the cooling tower or spray pond is handling the load, and if water distribution over the condensing surfaces is even.

Q   Should there be much of a temperature rise between the condenser outlet and the expansion valve?
A   No. The rise should be little as possible. To check this may not require a permanent thermometer in the liquid line ahead of each expansion valve, but there should be a thermometer well (2 in Fig. 14-1). Then readings can be taken if desired.

Q   What is the ideal temperature difference between the refrigerant and the substance being cooled?
A   This depends on several factors. Most important is the amount of surface in the evaporator; the greater the area, the less temperature difference needed. The insulating effect of ice and frost formed on direct-expansion coils and the speed of movement of the substance surrounding the evaporator also affect the temperature difference. The best results are attained in most plants when the difference amounts to about 10°F. Thermometer 3 (Fig. 14-1) in the space or substance to be cooled gives one part of the temperature difference and also gives direct indication of the effectiveness of the refrigerating process.

Q   Is it good practice to carry the saturated vapor all the way through the evaporator and into the suction line if the evaporator is close to the compressor?
A   No. Some superheating must be allowed, but since the superheated vapor is almost useless in heat removal, the amount of superheat must be kept down.

Q   What is known as the "stick finger test"? Is this good practice?
A   In small plants with one compressor and one evaporator the operator can become familiar with the best suction pressure to carry and the distance that the frost line should extend along the suction line to the compressor. Some operators scrape the frost from the evaporator outlet line and apply the stick finger test. If the finger sticks to the pipe, saturated refrigerant is passing through. If it doesn't, the refrigerant is leaving the evaporator in the superheated state. These methods are both crude. A better answer is to place thermometer 4 (Fig. 14-1) at the outlet of the evaporator. If there are many coils or other evaporators, place thermometer wells 5 (Fig. 14-1) at all important points where the branch suction lines connect into the main suction header.

Q   Why is a thermometer needed in the main suction line close to the compressor?
A   Thermometer 6 (Fig 14-1) gives information that helps to operate the compressor at greatest efficiency. If superheated gas passes into the compressor, it will be heated further by the hot valves, piston, and cylinder walls. This means less weight per cubic foot of gas and thus

reduces the actual amount of gas sent to the condenser. This in turn means less refrigerating effect for a given amount of compressor input.

Q   Upon what does the weight of the refrigerant sent to the condenser depend?

A   The weight of the refrigerant sent to the condenser depends on the density of the refrigerant. The higher the temperature, the lower the density. Since vapor is always at a higher temperature after entering the cylinder (60 to 65°F under usual operating conditions) than it is in the suction, any figures based on the latter would be misleading. A suction thermometer tells when the compressor suction valves become defective and leak. Any gas leaking through the suction valves will be highly super-heated. This will show plainly on a thermometer placed close to the compressor.

Q   What is the purpose of a thermometer in the discharge line?

A   Thermometer 7 (Fig. 14-1) in the discharge line from the compressor shows the heat of refrigerant leaving the cylinder on the way to the condenser. An operator can do little to control this temperature, other than see that refrigerant entering the compressor is at the right temperature. Also he should see that it's at the proper state of saturation or superheat and that the cylinder cooling water does its job of removing the heat of compression.

Q   Why is the discharge pressure gage so important in a refrigeration system?

A   A discharge pressure gage shows the pressure carried on the condenser, on the liquid receiver, and on the piping from the compressor through the condenser and receiver to the expansion valves. This gage gives warning when the pressure reaches a dangerous point. Reasons for this can be lack of cooling water, water that is too warm, or excessive amounts of noncondensable gases in the system. It also shows when condensing surfaces are scaled or dirty.

Q   Does the discharge pressure gage indicate the temperature of the liquid refrigerant leaving the condenser?

A   No. Pressure and temperature are not tied together for liquids as they are for saturated vapors. The closer the temperature of the liquid leaving the condenser comes to the temperature of the saturated vapor in the evaporator, the better will be the over-all efficiency.

Q   What does checking the temperature along the liquid line from the condenser tell the operator?

A   It tells him where insulation is needed. The receiver should be insulated if it is in a hot engine room. If it is outdoors, it should be protected from sun heat. The receiver should never be housed in a small shed or closetlike place without ventilation. If any part of the liquid line passes through a place where the room temperature runs higher than that of the liquid, the line should be covered. The reason is that there should be as little temperature rise as possible between the condenser outlet and the expansion valve.

Q In what size drums is ammonia shipped, and what precautions would you take in storing them?
A Dry (anhydrous) ammonia is shipped in steel cylinders holding 150, 100, or 50 lb net. Regardless of their size, they are filled so that at 65°F liquid ammonia fills only 88 per cent of the cylinder. The extra space allows for the expansion of ammonia at higher temperatures. Keep the cylinders away from steam lines, hot exhaust pipes, and heating devices. Never apply a torch to an ammonia cylinder.

Q On what part of the system must the charge connection be?
A The charge connection is always on the high-pressure side of the system. This prevents damage to the compressor by keeping out slugs of liquid. The connection is close to the receiver where the refrigerant may be stored.

Q How would you charge a large ammonia system?
A See Fig. 14-2. Most large plants take ammonia charge to the connection between the king valve and the expansion valve. First, raise the

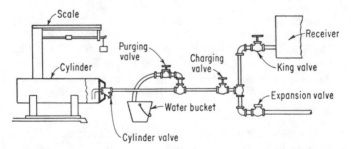

Fig. 14-2. Charging ammonia into large system between king valve and expansion valve.

bottom end of the cylinder about ¾ in.; be sure that the dipper pipe points down. If the charging line and connections are tight, close the king valve and open the cylinder valve and charging valve. After the king valve is closed, pressure between it and the expansion valve will drop to suction pressure (low side), allowing liquid ammonia to flow into the system. To be sure that the cylinder is discharging into the system, put it on the scale and watch the change in the weight. To charge a system with gas instead of ammonia, turn the cylinder over so that the inside dipper points up. Heating the cylinder with warm water will speed up the flow, but this is not recommended. Always check to make sure that the gas flows out of the cylinder and doesn't condense into it.

Q How much ammonia should a system contain?

A For the first charge or a complete recharge the best answer should come from the plant designer. If the figure isn't handy, you can calculate the amount needed from the data given in refrigerating handbooks. In a running plant the gage glass on the receiver tells the story; the receiver should be about one-half full. Without a gage glass you have to rely on watching the operation. If only the lower coils frost when the expansion valve is wide open and when the plant is not overloaded, the system needs ammonia. A clicking sound in the expansion valve also indicates a lack of ammonia.

Q Explain how you would withdraw ammonia from a system.

A Place the cylinder to be filled on the scale, and hook it up to the system as shown in Fig. 14-2 except that the dip tube should be pointed up. When everything is set, close the expansion valve and run the compressor slowly, keeping a full head of water on the condenser. Ammonia is thus pumped into the condenser, and liquid collects in the receiver. Watch the receiver gage glass to avoid overfilling. As soon as some liquid shows in the glass, open the cylinder valve and the receiver valve. Be sure the purging valve is closed. As ammonia enters the cylinder, watch the scale beam to check the rate of filling. After a time the flow will slow down owing to the pressure building up in the cylinder. Close the receiver valve and open the purging valve. This permits noncondensable gas to flow from the cylinder into the water, which will absorb ammonia gas coming with it. As long as bubbles show, noncondensable gas is present. Close the purging valve when the bubbling stops; then open the receiver valve and proceed with filling. When the scale beam shows that the right amount of ammonia has been added, making allowance for any ammonia that might have been in the cylinder at the start, close the receiver and cylinder valves and disconnect the cylinder.

Q What precautions should be taken when withdrawing ammonia from the system?

A Don't fill the cylinder above the 75 per cent level; it may explode if it gets heated. Start by weighing the cylinder with the cap off. Check this weight with the manufacturer's tare weight (on the cylinder tag) to see if the cylinder is empty. (ICC is strict with shippers who handle overfilled cylinders—they get a heavy fine and penalty.)

Q How would you charge a new Freon-12 refrigeration system?
A See Fig. 14-3. First, attach a compound gage to the suction gage connection. This gage shows pressure above atmosphere and inches of vacuum. Next, attach a pressure gage to the discharge connection. Place a Freon-12 service drum on the scale and connect it to the charging valve on the liquid line. Use flexible copper tubing or a flexible metallic hose to make this connection; solid piping puts a strain on the connec-

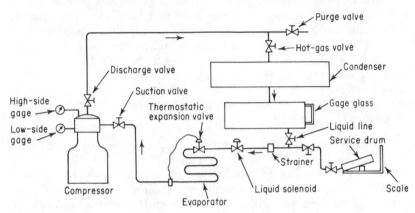

Fig. 14-3. Hookup used for charging a new Freon system.

tions as the scale lowers, causing leaks or breaks. Then, open the drum valve slightly to test the charging connections for leaks. Open the compressor discharge and suction valves, the condenser hot-gas and liquid valves, and the liquid solenoid valve. Then, open the charging valve and crack the drum valve enough to raise 40 psi in the entire system. Test every joint in the system with the exploring tube of a halide torch. If the system is for air conditioning, turn on the air-conditioning fan to put a load on the evaporator. Close the condenser liquid valve and note the weight of the service drum. Then, circulate water through the condenser and run the compressor intermittently until the system is charged. Freon should show in the bottom of the condenser gage glass. Close the charging valve and open the condenser liquid valve. For an air-conditioning system, keep the head pressures at 115 to 140 psi.

Q What precautions should be taken when charging or withdrawing gas?
A Be sure to ventilate the refrigeration machine room thoroughly. Start the ventilation system or fans and keep them running until the operation is complete. If there is no ventilating equipment, open all doors and windows.

Q   What happens when a cylinder is improperly hooked up for charging?

A   At times a mistake in hooking up the equipment, or some other error, may cause a reversed flow of ammonia (from system to cylinder). As a result the cylinder is overfilled. If the cylinder is exposed to even slight overheating, it may cause a dangerous explosion.

WARNING: Be careful when charging.

Q   What precaution would you take when starting a reciprocating compressor?

A   When a plant has been closed down for a time suction pressures may increase up to 50 psi in some ammonia systems, and liquid may accumulate at the low points in the suction line. Therefore, you must open the suction valves on the compressor slowly until the line is pumped down to normal working pressures.

Q   What effect does air have in a refrigeration system?

A   Air or other noncondensable gases cause excessive condensing pressure in a refrigeration system. To remove air, close the valve from the condenser receiver. Then, run the compressor until you are sure that most of the refrigerant charge has been pumped into the condenser receiver, but watch the high-pressure gage to be sure discharge pressure doesn't rise too high. With the compressor shut down, let the water flow through the condenser and note the outlet-water temperature. After an hour, when the cooling water should be balanced, observe the condenser pressure. Compare this pressure with the data on a chart that shows condenser-water temperature and corresponding condenser pressure for the refrigerant used. (All plants should have such a chart.) If the pressure is higher than it should be, purge the system by slowly opening the valve at the top of the condenser. Watch the pressure gage and also note the odor of the gas from the condenser to avoid loss of refrigerant. Be sure to cool the condenser until the gage pressure is as low as possible before purging. The larger systems have automatic purgers.

Q   Explain how you would keep the load balanced in a large refrigeration plant.

A   Large plants depend mostly on hand operation in starting and stopping compressors, pumps, etc. The operator must be alert in keeping the load balanced by running the right number of compressors. Maintaining correct temperatures in the boxes requires keeping the correct backpressures. This means starting up another compressor as the load rises; then as backpressures go down, one or more compressors must be stopped. In plants having only one compressor the load is balanced by cycling (starting and stopping). Smart operation calls for having machines running and ready to cut in before a large perishable load arrives—a carload of warm produce, for example. If the load is to be stored in a separate box, then the operator should have that box cut in and cool enough to attain proper temperatures as soon as possible after the load arrives. It is more economical and easier to keep temperatures from rising than to bring high temperatures down.

Q How would you purge a centrifugal unit?
A A built-in purge unit is usually fitted on a centrifugal system. Head pressures in these machines are less likely to give trouble than in reciprocating compressors. On most jobs it is common practice to run the purge unit before the main machine is started. This ensures ridding the machine of noncondensable gases.

Q What causes surging in centrifugal compressors?
A Surging in many centrifugals is normal and nothing to worry about, except at low loads. At 10 to 20 per cent of full load surging can cause the compressor to overheat, raising the bearing temperatures. So don't operate continuously under these conditions without getting advice from the manufacturer.

Q What must you watch carefully when operating an absorption or a steam-jet system?
A The steam supply is extremely important in both these systems, because without it no refrigerating effect can be produced. Be sure that there is always enough steam at the required pressure.

Q What harm is done by keeping suction pressure too low in a reciprocating system?
A If suction pressure is pulled too low, liquid refrigerant may reach the compressor suction where it can do damage. Watch suction-line frost; it is a fair indicator of whether the pressure is right.

Q List five important steps in starting an ammonia system.
A (1) Turn on the cooling water; (2) open the compressor discharge valve; (3) start the compressor; (4) open the compressor suction valve slightly until evaporator pressure is pumped down to about 20 psi, then open it wide; and (5) open the liquid valve and adjust the expansion valve to give the desired suction pressure.

Q List important steps in shutting down an ammonia system.
A (1) Close the liquid valve; (2) close the suction valve; (3) shut down the compressor; (4) close the compressor discharge valve; and (5) shut off the cooling water to the equipment.

Q Explain in detail what you would check at regular intervals after a new refrigeration system is started and running.

A   As the plant gets into regular operation, keep checking the bearings. cylinders, oil pumps, gages, thermometers, and other devices for signs of heating or other trouble. Adjust pressure-gage valves to keep the needles from vibrating excessively. Be sure the machine has enough oil in the crankcase; an extra inch or two is best for the first week of operation (new machine). Change the oil each week for the first month. Then open the crankcase and clean it before putting in fresh oil.

Test the automatic controls and safety cutouts. Adjust the float or expansion valves to balance the load; then put the system on automatic operation. At normal running conditions the receiver will be from one-fourth to one-half full of liquid refrigerant. The main liquid valve (king valve) extends into the receiver under the liquid's surface, even when the vessel is nearly empty. Have this valve wide open when the plant is running. Crack the expansion valve a part of a turn—just enough to keep the suction pressure at the point desired and to keep frost from accumulating along the suction line back to the compressor cylinders. However, frost on the suction trap does no harm.

Q   Explain how air and other noncondensable gases (foul gas) get into a refrigeration system.

A   There are four common sources of foul gas in the refrigeration system: (1) corrosion, (2) breakdown of oil, (3) decomposition of the refrigerant in contact with impurities, and (4) introduction of air under vacuum operation or during maintenance work.

Q   Why is it necessary to purge foul gas from the system?

A   While the refrigerant gases condense and decline in pressure and temperature, these foul gases add their pressure and temperature to the system, thus increasing the high-side pressure and also the power costs. The power wasted is very high when head pressures climb. For example, take an ammonia plant with a suction pressure of 15 psi and a head pressure of 215 psi, conditions which are not uncommon. This plant would use on the average of 1.85 bhp per ton of ice produced. If the head pressure is reduced to 185 psi by purging off the foul gases, the power needed is reduced to 1.66 bhp per ton of ice, or a saving of 11.45 per cent in power costs. This does not take into consideration the waste of refrigerant at the higher pressure or the added wear and tear on equipment. An efficient operator keeps foul gases purged from the system.

Q   What is an automatic purger?

A   Automatic purgers separate the ammonia from the foul gases and discharge them into the atmosphere. Various designs are on the market, but they all work on the same principle (see Fig. 14-4). To put a purger in operation, open the hand valve on the liquid line and suction line valve. The coil then reduces the temperature in the purger shell to near suction temperature. Open the valve in the gas line from the receiver, admitting refrigerant gas and foul gas mixture. The refrigerant condenses

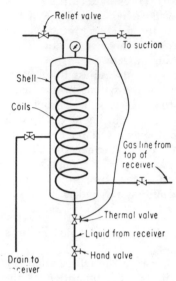

Fig. 14-4. Typical automatic refrigerator purger.

because of the low temperature, and foul gases collect in the upper portion of the shell. When sufficient foul gases have collected to raise the pressure above the relief valve pressure setting, the valve opens, venting gases to the atmosphere.

Q Explain how to hand-purge a system if there is no automatic purger installed.

A See Fig. 14-5. Run the compressor long enough to pump all the air and noncondensable gases into the high side; then shut it down. Run water over the condenser until it is cool, thus condensing all the refrigerant. Crack the purge vent valve in the highest part of system above the condenser so it expels air and noncondensable gases. If there is no such vent valve, be sure to install one.

CAUTION: Always wear goggles when working with a refrigerant.

Some older low-pressure plants have a hose from this vent leading into a bucket of water to absorb the ammonia odors. The vent is cracked each watch. However, the most efficient method is an automatic purger.

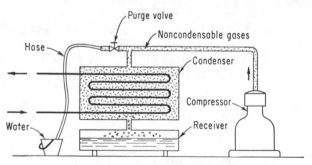

Fig. 14-5. Hand purging can be done with hookup like this.

Q How can you tell if a Freon-12 system needs purging?

A (1) Close the liquid-line valve at the receiver and pump the low side down to about 6 to 10 psi. (2) Stop the compressor, but let the condenser water continue to flow until the temperature of the liquid in the condenser receiver and in the condenser remains constant. (3) Determine the temperature of the liquid at the bottom of the receiver. This temperature will be an approximation of the condensing temperature. (4) Read the condensing temperature that corresponds to the condensing pressure by the high-pressure gage. (5) Subtract the existing condensing temperature from the temperature of pure Freon-12 at the existing condenser pressure. If the difference between these two temperatures is more than 5°F, the system must be purged.

Q   What type of purger is used on a centrifugal compressor system?
A   Either a thermal or a compressor type. The purpose of the unit is to evacuate air and water from the centrifugal machine and to recover and return refrigerant which is mixed with the air. Even though a machine may be perfectly airtight, it may develop a water leak which is detected only by operation of the purge system. If water is allowed to remain in the machine, serious damage to tubes and other internal parts can result.

The compressor-type purge unit operates independently, using a small reciprocating compressor to remove the air, moisture, and a small quantity of refrigerant from the condenser. The thermal-type purge unit operates on a pressure differential principle, and a reciprocating compressor is not needed.

Q   Name the two main causes of expansion valve malfunction and explain how to repair them.
A   (1) The power assembly loses charge, and (2) moisture freezes in the valve. The valve closes if it loses its charge. If this happens, replace the power assembly and feeler bulb in large valves, or replace the entire valve on smaller installations.

Moisture may freeze the valve shut, restricting or stopping refrigerant flow. This starves the evaporator and causes low suction pressure. If the valve freezes open, the system floods and suction pressure is high. Ice, caused by moisture in the system, can form whenever the evaporator operates at below-freezing temperatures.

When the compressor stops and the expansion valve and system warm up, operation is okay until ice forms again. This may cause the compressor to short-cycle. To melt ice, wrap the valve with towels soaked in hot water. Then install a new dryer in the line feeding the expansion valve. Operate the system above freezing for a few days. Moisture should then be trapped in the dryer. If moisture persists, evacuate the refrigerant charge and oil, and dry the system with a vacuum and heat, or blow dry CO_2 through it.

Improper pressures or superheat may also indicate other problems.

See the manufacturer's service manuals for details on correcting expansion-valve trouble.

Q When two or more centrifugal refrigeration units of different size are on the same load, how is the output balanced for best economy?
A Adjust brine and water flow rates through each unit so they are proportional to its full-load rating.

> EXAMPLE: A plant has one 300-ton and one 500-ton centrifugal refrigeration unit. Total brine flow is 1,600 gpm, and condenser-water flow is 2,000 gpm. How much brine and water must flow through each machine for an 800-ton load?
>
> SOLUTION: Brine flow, in gpm per ton, is found by dividing total flow by total plant capacity, or $1,600/800 = 2$ gpm per ton. Also, condenser flow is $2,000/800 = 2.5$ gpm per ton. Flow rates should then be:
>
> Brine through small machine = 300 tons × 2 gpm per ton = 600 gpm
> Brine through large machine = $500 \times 2 = 1,000$ gpm
>
> Condenser-water flow is $300 \times 2.5 = 750$ gpm for the small unit, $500 \times 2.5 = 1,250$ gpm for the large unit.

Q Does low-temperature operation (below $-40°F$) cause additional problems with the compressor and the system?
A Yes, breakdowns are common when the system is operating below $-40°F$, as the following statement from the ASHRAE Data Book points out: "In general, production below $-40°F$ requires a special technique, somewhat different from that commonly used with ordinary refrigeration compressors."

These seven operating techniques must be followed: (1) maintain the best possible condensing medium; (2) be sure to eliminate all noncondensable gasses; (3) keep the suction gas, both from the evaporator and from between the stages in multistage systems, as cool as possible; (4) have the best possible valve plate and compressor efficiency; (5) use multistage compressors; (6) have a good oil separator; and (7) if all else fails, slow down. Use a larger unit if capacity is too low.

MAINTENANCE

Q Name the five common methods of testing for refrigerant leaks, and describe each.
A (1) Sulfur candle test, (2) halide torch test, (3) leak-tracing dye or harmless odorant test, (4) soap and water test, and (5) ammonia swab test.

The sulfur candle test is used only for ammonia leaks. A leak is indicated by the cloud of white smoke generated when the fumes from the sulfur candle come in contact with escaping ammonia.

The halide torch test is for finding leaks of halocarbon refrigerants only. The open flame must be used carefully when refrigerants of this type are being tested. If there is a fluorine-compounded gas in the air being tested, the high-temperature flame of the torch causes the refrigerant to break down and form a volatile halide. The flame then changes to blue or bright green when the air contains as little as 0.01 per cent of Freon.

BEWARE: Don't breathe the fumes of this torch because on contact with high temperatures refrigerant breaks down to a gas containing phosgene. The gas is also semitoxic in small quantities and deadly in larger volumes. If you think there is enough contamination to form a flammable mixture when systems containing methyl chloride, methylene chloride, ethyl chloride, or bichlorethylene are to be tested, use the soap and water test.

Leak-tracing dye and some harmless odorant are often introduced into a system to find leaks.

Swabbing suspected parts with a soap and water solution is another good way to locate leaks. It is used mostly on systems containing carbon dioxide and the highly flammable refrigerants such as ethane, propane, butane, and isobutane.

The ammonia swab test is used for systems containing sulfur dioxide.

Q   Describe a halide torch, and explain how it is used.

A   A halide torch is a sensitive probe for detecting Freon leaks as small as 0.01 per cent by volume. It is a better and faster method of detection than using soap and water around suspected parts. The torch has a replaceable copper reaction plate (disk) surrounded by a metal shield that has a window through which you can watch the flame. The body of the torch has a gas passage and intake to admit air for burning. A soft-rubber detector tube is attached to the air intake. The free end of the detector tube is passed over suspected joints. The torch is fed from a small portable tank of acetylene gas. When the torch burns properly, a draft of 3 ft per sec is induced through the rubber tube. If the open end is passed over the leak, fumes are drawn in and the color of the flame changes. Normally, the flame is colorless and somewhat pear-shaped. A very slight halocarbon leak fringes the colorless flame with green. As the leak gets stronger, the entire flame turns first to a bright green and finally to a bright green fringed with red. Serious leaks snuff out the torch flame by cutting off its air supply.

REMEMBER: A halide torch is not reliable if the air in the space is contaminated with refrigerant.

Q   How would you test safety valves in an ammonia refrigeration sys·
tem?

A   See Fig. 14-6. The simplest way to test safety valves is during shut·
down. First, replace the safety valves. Then, hook up the used valves to

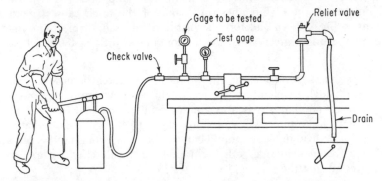

Fig. 14-6. Testing safety valve and gages is important; must be done periodically.

the fittings. Use fittings of extraheavy steel of 500 psi. Connect a common test gage, graduated up to 600 psi. With this arrangement either
a pressure gage or a relief valve may be tested separately. Fill a high-
pressure grease gun with a medium-heavy-grade lubricating oil. Pump
the pressure up with the grease and check the gage. After testing the
pressure gage, close the gage valve and test the relief valve. Drain the
overflow from the valve into a can. Check all the relief valves in your
refrigeration system periodically. It is the only way to be sure that they
work. If the discharge line from the relief valve is piped outside of the
building instead of back to the suction line, check the relief piping to
make sure it is clear.

Q   Why is moisture harmful in a Freon system?

A   Water vapor in a Freon system freezes the automatic regulating
valves. Moisture also causes oil breakdown and sludging, gummed-up
bearings, and similar troubles.

Q   Name some of the more common driers.

A   Activated alumina is a granular aluminum oxide that removes moisture by absorption. Silica gel is a glasslike silicon dioxide which also
removes moisture by absorption. Drierite is an anhydrous calcium sulfate
prepared as a granular white solid. It removes moisture by chemical action. Calcium oxide and calcium chloride remove water and acid by
chemical action.

Q   How are driers used?

A   Arrange pipelines so that the drier can be connected temporarily into the system and Freon circulated through it for about 12 hr. An alumina drying material usually does the job in that time. If it is left in the system, a fine white powder, representing dry material that has slaked out, will go through the system and cause trouble. The powder then passes through liquid strainers and collects into puttylike messes at the expansion valves.

Q   How would you remove a cylinder head from a large vertical ammonia compressor?

A   Before removing the cylinder head, make sure the pressure has been relieved. Then look at the long studs (see Chap. 4, Fig. 4-7). Remove the nuts on the short studs first, then back off the nuts on the long studs a few turns. See if the springs holding the safety head begin to lift the cylinder head.

> BEWARE: Never remove all the nuts from the long studs until the joint is broken. If all the nuts are removed while the cylinder head sticks to gasket, there is enough spring tension in large units to throw the head off when the joint is suddenly released. This can cause serious injuries.

Mark the safety heads for position and remove them.

> CAUTION: Don't damage the head's ground joint.

Next remove the piston and connecting rod. If the unit is large and the crank bearings are heavy, place a jack under the bottom half of the connecting-rod bearing to hold it in position while dismantling. Use only safe lifting gear. Always screw eyebolts down tight to the bottom of tapped threads. Tie the eyebolt to the hoisting hook so it can't jump out when the piston is lowered.

Q   When opening up a refrigeration compressor, how would you check the cylinders and piston rings?

A   Looking down on a compressor, cylinder wear is usually greatest on the sides at right angles to shaft center line in the piston-ring belt. Take micrometer readings and jot down these measurements. The cylinders may be slightly out-of-round or tapered. If they are not worn enough for reboring, replace the common snap rings with newer seal-joint or two-piece seal rings. On small units, under 6-in. bore, 0.0015 in. side clearance on the ring groove is about right. Allow 0.002 to 0.003 in. on larger pistons. Try the new rings in the smallest diameter of the cylinder for gap clearances before placing them on piston.

Q   Where would you look for wear on trunk pistons?

A   A trunk piston acts as a crosshead and eventually wears out-of-round. Badly worn pistons up to 4 in. in diameter are cheaper to replace than to repair. Repair larger units by turning down the piston below the ring belt. Then build up the metal by spraying on a high-grade bronze, and machine it to the original diameter.

Q   Insurance figures show that more refrigeration compressors fail from faulty discharge valves than from any other cause. Explain how you would check these valves.

A   Valves should be inspected at least once a year. Flaws can be detected on large disks by suspending the disk from a cord and striking a light hammer blow. If the disk is sound and the stem is free of cracks where it is welded to the disk, it will usually ring with a clear, bell-like sound. Newer machines have small plate-type suction valves. They are easier to remove and repair. Don't use valves that look fairly good. They must be perfect. A few leaky valves cost more in refrigeration and power loss in a short time than an entire new set of valves. The ground joint between the cylinder and safety head must be tight. Polish the joint with a very fine compound. In grinding, keep rotating the head completely around rather than back and forth in half or quarter turns. On some larger valves the bearing surface has a slight radius. This seat must be kept narrow. To grind, use a fine compound and finish with light machine oil. Grinding the valve ring on a magnetic chuck in shop does the best job. If the valves have a dashpot in the suction-valve cage, see that the vents are clear. A spring supports the weight of the valve disk and stem. Adjust the spring just enough to hold the valve slightly off its seat. This helps the valve to open as soon as the pressure in the cylinder and suction port equalize. Then inertia closes the valve at the suction stroke's end. These valves are hard to remove, and their stems sometimes fail. Check stems, threads, and nuts carefully. If the threads are dressed, avoid sharp tool cuts on the valve stem.

Q   How is moisture removed from a new system?

A   As soon as the new piping system has been leak-tested, pull a vacuum of 0.2 in. absolute pressure (29.8 in. of vacuum) in the system with an ambient temperature above 40°F. Use a high-vacuum pump. When this pressure is reached, all free moisture will have boiled into water vapor. More important, the moisture will have been pumped out by a pump that creates 2 mm absolute pressure for dehydrating. A vacuum indicator is also needed. Fig. 14-7 shows the hookup of pump and vacuum indicator.

> NOTE: If the system has been flooded from a condenser leak, you have to break low points in the lines and drain out water before pumping.

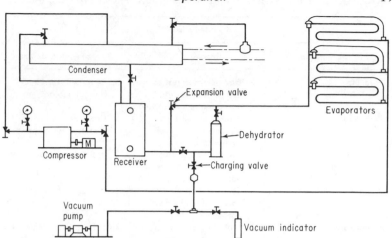

Fig. 14-7. Vacuum pump and indicator are hooked into charging line.

Q What are the most common accidents with reciprocating compressors?

A Of 150 accidents reported by the Hartford Steam Boiler Inspection and Insurance Co., 65 accidents were caused by a valve, valve cage, etc.; 36 by bearings; 14 by a piston or rings; 11 by the cylinder or head; 7 by the shaft; and 17 by other problems. Valves failed because of cracks, with liquid carry-over being the major cause. Bearings failed mostly because of neglect of the oil level or dirt in pipe lines.

TROUBLESHOOTER'S GUIDE TO REFRIGERATION PROBLEMS

Trouble: Compressor Fails to Start		
Probable cause	Symptoms	Recommended action
1. Power failure	1. Electric circuit test shows no current on line side of motor starter	1. Check for blown line fuse or broken lead
2. Disconnect switch open	2. Electric circuit test shows no current on line side of motor starter	2. Determine why switch was opened. If everything OK, close switch

Trouble: Compressor Fails to Start (contd)

Probable cause	Symptoms	Recommended action
3. Fuse blown	3. Electric circuit test shows current on line side but not on motor side of fuse	3. Replace fuse. Check load on motor
4. Low voltage	4. Electric circuit tester glows but not at full brilliance	4. Check with voltmeter, then call power company
5. Burned-out motor	5. Full voltage at motor terminals but motor will not run	5. Repair or replace
6. Inoperative motor starter	6. Test for burned-out holding coil or broken contacts	6. Repair or replace
7. Open control circuit a. dual pressure control b. oil failure control c. motor starter thermal overloads d. thermostat not set for cooling e. open circuit from "interlocking" relays	7. Motor starter holding coil is not energized	7. Locate open control and determine cause. See individual control instructions
8. Broken or sheared coupling	8. Motor runs but compressor does not	8. Repair or replace. Properly realign

Note: See Chap. 8 for more electrical troubleshooting information.

9. Frozen compressor due to locked or damaged mechanism	9. Compressor will not operate	9. Overhaul compressor
10. Suction pressure below cutin setting of low-pressure cutout switch	10. Open contacts on low-pressure switch. Suction pressure below cutin setting	10. Check for loss of refrigerant. Repair leak and recharge
11. Discharge pressure above cutin setting of high-pressure cutout switch	11. Open contacts on high-pressure switch. Discharge pressure above cutin setting	11. Check condenser cooling water, fouled condenser, overcharged system
12. Oil pressure failure control switch has cutout	12. System will restart by resetting oil pressure failure control switch	12. Check oil level, oil pressure, wiring, and control for faulty control

Trouble: Compressor "Short-cycles"

1. Intermittent contact in electric control circuit	1. Normal operation except too frequent stopping and starting	1. Repair or replace faulty electrical control

Trouble: Compressor "Short-cycles" (contd)

Probable cause	Symptoms	Recommended action
2. Low-pressure controller differential set too close	2. Normal operation except too frequent stopping and starting	2. Reset differential in accordance with proper job conditions
3. Leaky liquid line solenoid valve	3. Valve may hiss when closed. Also temperature change in refrigerant line through valve	3. Repair or replace
4. Dirty or iced evaporator	4. Reduced air flow a. dirty air filters b. broken fan belt c. fan belt tension improperly adjusted	4. Clean or defrost evaporator. Check filters and fan drive
5. Faulty condensing	5. Excessively high discharge pressure	5. Check for water failure or evaporative condenser trouble
6. Overcharge of refrigerant or noncondensable gas	6. High discharge pressure	6. Remove excess refrigerant or purge noncondensable gas
7. Lack of refrigerant	7. Normal operation except too frequent stopping and starting on low-pressure control switch	7. Repair refrigerant leak and recharge
8. Water regulating valve inoperative or restricted by dirt, or water temperature too high	8. High discharge pressure	8. Clean or repair water valve
9. Water piping restricted or supply water pressure too low	9. High discharge pressure	9. Determine cause and correct
10. Restricted liquid line strainer	10. Suction pressure too low and frosting at strainer	10. Clean strainer
11. Faulty motor	11. Motor starts and stops rapidly	11. Repair or replace faulty motor
12. Fouled shell-and-tube condenser	12. Compressor cuts off on high-pressure cutout	12. Clean condenser tubes
13. Faulty operation of evaporative condenser	13. Compressor cuts off on high-pressure cutout a. no water b. spray nozzles clogged c. water pump not operating d. coil surface dirty e. air inlet or outlet obstructed f. fan not operating	13. Determine cause and correct a. fill with water b. clean spray nozzles c. repair faulty pump d. clean coil e. remove obstruction f. repair

Trouble: Compressor Runs Continuously

Probable cause	Symptoms	Recommended action
1. Excessive load	1. High temperature in conditioned area	1. Check for excessive fresh air or infiltration. Check for inadequate insulation of space
2. Thermostat controlling at too low a temperature	2. Low temperature in conditioned area	2. Reset or repair thermostat
3. "Welded" contacts on electrical control in motor starter circuit	3. Low temperature in conditioned space	3. Repair or replace faulty control
4. Lack of refrigerant	4. Bubbles in sightglass	4. Repair leak and charge
5. Overcharge of refrigerant	5. High discharge pressure	5. Purge or remove excess
6. Leaky valves in compressor	6. Compressor noisy or operating at abnormally low discharge pressure or abnormally high suction pressure	6. Overhaul compressor
7. Solenoid stop valve stuck open or held open by manual lift stem	7. Air-conditioned space too cold	7. Repair valve or restore to automatic operation

Trouble: Compressor Loses Oil

Probable cause	Symptoms	Recommended action
1. Insufficient oil charge	1. Oil level too low	1. Add sufficient amount of proper compressor oil
2. Clogged strainers or valves	2. Oil level gradually drops	2. Clean or repair and replace
3. Loose expansion valve or remote bulb	3. Excessively cold suction	3. Provide good contact between remote bulb and suction line
4. Liquid flooding back to compressor	4. Excessively cold suction. Noisy compressor operation	4. Readjust superheat setting or check remote bulb contact
5. Short-cycling	5. Too frequent starting and stopping of compressor	5. Defrost; check pressure cutout
6. Crankcase fittings leak oil	6. Oil around compressor base and low crankcase oil level	6. Repair oil leak and add proper refrigerant oil

Trouble: Compressor Is Noisy

Probable cause	Symptoms	Recommended action
1. Loose compressor drive coupling	1. Coupling bolts loose	1. Tighten coupling and check alignment
2. Lack of oil	2. Compressor cuts out on oil failure control	2. Add oil
3. Dry or scored seal	3. Squeak or squeal when compressor runs	3. Check oil level

194

TROUBLESHOOTER'S GUIDE TO REFRIGERATION PROBLEMS (Continued)

Trouble: Compressor Is Noisy (contd)

Probable cause	Symptoms	Recommended action
4. Internal parts of compressor broken	4. Compressor knocks	4. Overhaul compressor
5. Liquid "flood back"	5. Abnormally cold suction line. Compressor knocks	5. Check and adjust superheat. Valve may be too large or remote bulb loose on suction line. Air entering evaporator too cold for complete evaporation of liquid.
6. Dirty water-regulating valve, too high water pressure or intermittent water pressure	6. Water valve chatters or hammers	6. Clean water-regulating valve. Install air chamber ahead of valve
7. Expansion valve stuck in open position	7. Abnormally cold suction line. Compressor knocks	7. Repair or replace
8. Compressor or motor loose on base	8. Compressor or motor jumps on base	8. Tighten motor or compressor hold-down bolts

Trouble: System Short of Capacity

Probable cause	Symptoms	Recommended action
1. Flash gas in liquid line	1. Expansion valve hisses	1. Add refrigerant
2. Clogged strainer or solenoid stop valve	2. Temperature change in refrigerant line through strainer or solenoid stop valve	2. Clean or replace
3. Ice or dirt on evaporator	3. Reduced air flow	3. Clean coil or defrost
4. Expansion valve stuck or obstructed	4. Short-cycling or continuous running	4. Repair or replace expansion valve
5. Excess pressure drop in evaporator	5. Superheat too high	5. Check superheat and reset thermostatic expansion valve
6. Improper superheat adjustment	6. Short-cycling or continuous running	6. Adjust expansion valve. Check superheat and reset thermostatic expansion valve
7. Expansion valve improperly sized	7. Short-cycling or continuous running	7. Replace with correct valve

Trouble: Discharge Pressure Too High

Probable cause	Symptoms	Recommended action
1. Too little or too warm condenser water	1. Excessively warm water leaving condenser	1. Provide adequate cool water, adjust water-regulating valve
2. Fouled tubes in shell-and tube condenser	2. Excessively cool water leaving condenser	2. Clean tubes

Trouble: Discharge Pressure Too High (contd)

Probable cause	Symptoms	Recommended action
3. Improper operation of evaporative condenser	3. Low air or spray water volume. Scaled surface	3. Correct air or water flow. Clean coil surface
4. Air or noncondensable gas in system	4. Exceptionally hot condenser and excessive discharge pressure	4. Purge
5. Overcharge of refrigerant	5. Exceptionally hot condenser and excessive discharge pressure	5. Remove excess or purge

Trouble: Discharge Pressure Too Low

Probable cause	Symptoms	Recommended action
1. Too much condenser water	1. Excessively cold water leaving condenser	1. Adjust water-regulating valve
2. Lack of refrigerant	2. Bubbles in sightglass	2. Repair leak and charge
3. Broken or leaky compressor discharge valves	3. Suction pressure rises faster than 5 lb/min after pressure shutdown	3. Remove head, examine valves, replace faulty ones
4. Leaky relief bypass valve	4. Low discharge pressure and high suction pressure	4. Inspect valve to determine if replacement is necessary

Trouble: Suction Pressure Too High

Probable cause	Symptoms	Recommended action
1. Excessive load on evaporator	1. Compressor runs continuously	1. Check for excessive fresh air or infiltration, poor insulation of spaces
2. Overfeeding of expansion valve	2. Abnormally cold suction line. Liquid flooding to compressor	2. Regulate superheat setting expansion valve, see remote bulb OK on suction line
3. Expansion valve stuck open	3. Abnormally cold suction line. Liquid flooding to compressor	3. Repair or replace valve
4. Expansion valve too large	4. Abnormally cold suction line. Liquid flooding to compressor	4. Check valve rating, replace if necessary
5. Broken suction valves in compressor	5. Noisy compressor	5. Remove head, examine valves, repair faulty ones

Trouble: Suction Pressure Too Low

Probable cause	Symptoms	Recommended action
1. Lack of refrigerant	1. Bubbles in sightglass	1. Repair leak, then charge system
2. Light load on evaporator	2. Compressor short-cycles	2. Not enough refrigerant
3. Clogged liquid-line strainer	3. Temp. change in refrigerant line through strainer or solenoid stop valve	3. Clean strainer

TROUBLESHOOTER'S GUIDE TO REFRIGERATION PROBLEMS (Continued)

Trouble: Suction Pressure Too Low (contd)

Probable cause	Symptoms	Recommended action
4. Expansion valve power assembly has lost charge	4. No flow of refrigerant through valve	4. Replace expansion valve power assembly
5. Obstructed expansion valve	5. Loss of capacity	5. Clean valve or replace if necessary
6. Contacts on control thermostat stuck on closed position	6. Conditioned space too cold	6. Repair thermostat or replace if necessary
7. Compressor capacity control range set too low	7. Compressor short-cycles	7. Reset compressor capacity control range
8. Expansion valve too small	8. Lack of capacity	8. Check valve rating table for correct sizing and replace if necessary
9. Too much pressure drop through evaporator	9. Too high superheat	9. Check for plugged external equalizer

SUGGESTED READING

Reprints sold by *Power* magazine, costing $1.20 or less

    Power Handbook, 64 pages
    Air Conditioning, 20 pages
    Refrigeration, 20 pages
    Handbook on Fans, 16 pages
    Balancing Rotating Equipment, 24 pages
    Vibration Isolation, 16 pages

Books

    Elonka, Steve, and Joseph F. Robinson: *Standard Plant Operator's Questions and Answers*, vols. I and II, McGraw-Hill Book Company, New York, 1959.

    Elonka, Steve: *Plant Operators' Manual*, rev. ed., McGraw-Hill Book Company, New York, 1965.

    ASHRAE Data Book, American Society of Heating, Refrigerating and Air-Conditioning Engineers.

    Emerick, Robert H.: *Troubleshooters' Handbook for Mechanical Systems*, McGraw-Hill Book Company, New York, 1969.

15

REFRIGERATION APPLICATIONS

It would be impossible to cover in one chapter, or even in one book, all the applications of refrigeration in our modern way of life. The list is virtually endless, since new applications are developed each day. The intent of this chapter, of necessity, will be to outline some of the better-known uses of refrigeration and to list some of the newer applications or fields of application. For further details we suggest the reader refer to trade publications, reprints of specific articles, or other books dealing with a particular field.

Q What industry accounts for the greatest percentage of refrigeration capacity?
A Before the advent of dependable mechanical refrigeration nearly all the means of maintaining a temperature in a given space lower than that of the surrounding space were for the preservation of food. Though mechanical refrigeration has increased the ability to control temperatures and conditions to such a degree that it is used in nearly all industries, the preservation of food still accounts for more refrigeration capacity and dollar investment than any other industry.

Q In general, what principle of low or controlled temperatures makes refrigeration so important to the food industry?
A All natural foods, i.e., fruits, vegetables, meats, and dairy products, are living organisms. Even after these foods are removed from their natural surroundings to be prepared for human consumption, they continue to "live," and certain changes occur within the product. These changes may be physical, chemical, or a combination of the two. Physical changes include crystallization, expansion, and dehydration or desiccation. Chemical changes include ordinary chemical changes, enzymic changes, and those caused by microorganisms (bacteria, yeasts, and molds). Most of these changes occur more rapidly at high temperatures, so, in general, control of temperature and humidity conditions makes it possible to control the rate of change. Heat is required to create and develop life, but if it is not controlled, it can hasten the destruction of

198

living matter. Low temperature retards the rate of destruction and can be utilized to maintain foods for long periods.

Q Does cold storage or freezing improve foods?

A Except for some foods that require controlled temperatures as a part of an "aging or mellowing" process, no food product is improved by cold storage. One cannot expect to put poor-quality products into cold storage and remove high-quality products at a later date.

Q What temperature ranges are used in the preservation and storage of perishable foods?

A Foods that are intended for consumption within a few days or weeks are usually stored at temperatures a few degrees above their freezing point (+28 to +40°F). If storage is for a longer period, products are usually frozen by one of several methods.

Q What is quick freezing?

A In commercial terms quick freezing is any one of several refrigeration processes in which a product is frozen so rapidly that it is not greatly changed by freezing. One authority defines quick freezing as follows: "The zone of maximum crystal formation, which means solidification, must be passed through in 30 minutes or less. Such speed insures small crystals and a minimum disturbance of tissue structure."

Q What is a sharp freezer?

A A sharp freezer is a well-insulated room maintained at a low temperature. Heat from the product is absorbed by the air and transferred to the refrigeration coils by natural convection currents.

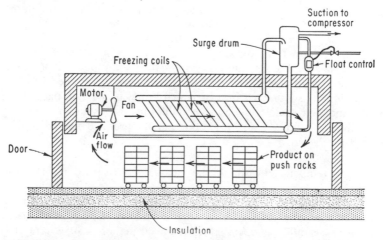

Fig. 15-1. Blast freezer maintains high velocity of low-temperature air for fast freezing.

Q What is a blast freezer?

A A blast freezer is any system where low air temperatures are maintained and blowers or fans direct air over the product at high velocities, usually from 500 to 1,500 fpm (see Fig. 15-1). A typical blast freezer is the tunnel type, in which the refrigerant coils are located overhead and fans circulate the air over the product. The product may be moved through the tunnel on racks or by a conveyor.

Q What is contact freezing?

A Early contact freezers were shelf coils made up of pipes placed close together. Cold brine or refrigerant was circulated through the pipes, and the product to be frozen was placed on the coils. Later developments brought about the plate coil, which provided a more uniform surface for freezing. Present contact freezers make use of a number of horizontal plate coils (see Fig. 15-2) which are smooth on both sides. The space between the plates can be varied by the use of proper spacers. An electrically operated hydraulic piston compresses the packages of the product between the plates. This ensures firm contact on both sides of the package for fast heat transfer. Another type of contact freezer employs movable vertical plates, with the product entering at the top and moving through to the bottom.

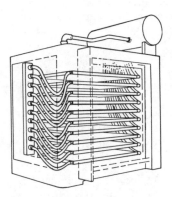

Fig. 15-2. Plate freezer removes heat by actual contact with package.

Q What is immersion freezing?

A Immersion freezing is the process in which the product is submerged in or sprayed with a refrigerated liquid. This has been applied to canned fruits and juices submerged in refrigerated brines or alcohol, bagged poultry in brine, and some fruits and vegetables in sugar solutions. The cold liquid may be pumped over the product, or the product may be conveyed through the liquid.

Q What temperatures should be maintained for the most efficient freezing?

A Some immersion freezers provide the desired results when using liquids at temperatures as high as 5°F. For best results, air-blast and contact freezers should be operated at temperatures of −30°F and lower. Experimental work in progress is using liquid nitrogen at −320°F.

Q What temperatures should be maintained for the best storage of frozen foods?

A Temperatures of rooms for frozen-food storage should never be above 0°F. Some meat products and some prepared foods require even lower temperatures. Research has proved that quality is maintained at a high level for a longer period if storage temperatures are maintained at −10°F and lower. Another very important factor is to maintain a constant storage temperature. Fluctuating temperatures have a greater detrimental effect on food vitamins, color, and texture than any other cause after the product has been frozen.

Q What methods are used to maintain high humidity in food storage rooms?

A The real cause of low humidity in cold-storage spaces is too great a temperature difference between the room temperature and coil surfaces. The moisture in the air will deposit on coil surfaces that are colder than room temperature. The lower the temperature of the coil surface, the lower the dew point of the air in contact with the coils. Moisture may be added to the air by the use of steam jets, sprays, humidifiers, etc., but if low-temperature coil surfaces are required to maintain room temperature, this additional moisture will be deposited on the coil surfaces as frost. To ensure high humidity, sufficient coil surface must be provided to maintain room temperature with a minimum temperature difference between the refrigerant and room air.

Q Why is a good refrigeration system so important to a meat-packing plant?

A Freshly slaughtered and dressed carcasses must be cooled quickly in order to retard bacterial action, which causes deterioration and loss of product. Equipment must be properly engineered to remove the heat from the carcasses without excessive air circulation. Humidity must be held at a reasonably high point, usually 85 per cent or above, to prevent excessive shrinkage and loss of color.

Q What changes in refrigeration systems for meat plants have been developed during the past few years?

A Until a few years ago most chill rooms were constructed with a brine spray chamber located over the meat rails (see Fig. 15-3). Air circulation was by natural circulation aided by the action of the sprays. Brine was chilled in a remote evaporator and pumped to the sprays, returning by gravity. Later developments added pipe coils in the spray deck space for direct expansion of the refrigerant. The spray served to keep the coils defrosted at all times and aided in maintaining high humidity.

Newer plants have been equipped with unit coolers of various types (see Fig. 15-4). Floor-mounted unit coolers with air ducts leading to the space above the meat rails have been used, both in brine spray and

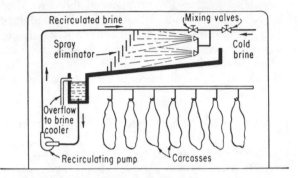

Fig. 15-3. Brine spray absorbs heat from air to cool meat.

periodic defrost units. Where sufficient space is available above the meat rails, small units are placed to provide sufficient refrigeration capacity and air distribution. Whatever the equipment used, care should be taken

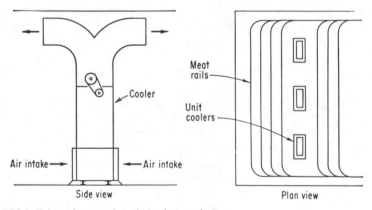

Fig. 15-4. Unit coolers use forced circulation of air.

to provide enough capacity to perform the necessary cooling without causing excessive shrinkage due to low humidity.

Q What range of temperatures are to be found in meat-packing plants?
A For rooms where freshly killed carcasses are to be cooled, temperature should be maintained at 30 to 34°F, with a maximum temperature of 38°F during peak loads. Holding coolers should be maintained at 32

to 34°F. Pork cutting rooms are maintained at 45 to 50°F with relative humidity low enough to prevent condensation of moisture on the carcasses entering the cutting room from the cooler. Beef, veal, and lamb cutting and boning rooms are held at 36 to 38°F. Smoked and cured meats are held at 28 to 32°F. Bacon for slicing is usually tempered by storing it at 24 to 28°F for 12 to 24 hr before slicing.

Q What is meant by precooling fruits?

A When fruit is harvested, the internal (or pulp) temperature is many degrees above the most desirable storage temperature. It is desirable to remove this internal or "field" heat as rapidly as possible to maintain high quality. When the fruit is harvested at some distance from a properly equipped cold-storage plant, portable refrigerating equipment is sometimes set up in the field to cool the fruit. It is common practice to circulate large volumes of air rapidly over and through the fruit to remove the field heat. This is the precooling cycle. Once the fruit is reduced to storage temperature, the volume of air is reduced to storage conditions.

Q What is heat of respiration?

A Sometimes called "heat of evolution," the heat of respiration is the heat created by internal changes in a fresh product while in storage. The rate per ton per hour increases as the storage temperature increases, and in the case of some vegetables, notably green beans and asparagus, it can become greater than the capacity of the refrigeration system to remove it.

Q What is "controlled atmosphere storage," and how does it affect fruit storage?

A The respiration of fruit in storage has a tendency to use a portion of the oxygen content of the air in the storage room. The fruit also gives off various gases such as carbon dioxide and ethylene. In some cases these gases have a detrimental effect on the fruit in storage, so they must be removed from the room. Tests have shown that certain fruits have a better storage life in an atmosphere with a low oxygen and high carbon dioxide content. By using spray-type air washers with proper concentrations of sodium hydroxide, the CO_2 content is controlled with greatest efficiency. Controlled atmosphere storage is also being tested in connection with meat storage.

Q How is artificial ice used in transporting fresh fruit and vegetables?

A Special railroad cars were developed many years ago for the shipment of fresh fruit, meat, and vegetables under refrigeration. These cars are well insulated and have spaces at each end where ice can be loaded from the top of the car. Floor racks are installed to provide air circulation around the product. Fans are mounted in the end of the cars and operated by friction drive from the car wheels when the car is in motion. By keeping proper ice levels in the bunkers, proper storage conditions are maintained throughout the period of shipment.

Q   How is artificial ice made?
A   In plate ice plants (now considered obsolete) blocks of ice 10 to 14 in. thick, 8 to 12 ft wide, and 10 to 20 ft long were frozen on refrigerated plates suspended in a tank of fresh water. These plates were made of continuous pipe coils between light steel plates. When the ice had reached the proper thickness, it was loosened from the plates by circulating warm brine or warm ammonia. The ice was then transported to a table where it was sawed up into smaller pieces for handling. Plate ice plants have been superseded by raw water can plants (see Fig. 15-5)

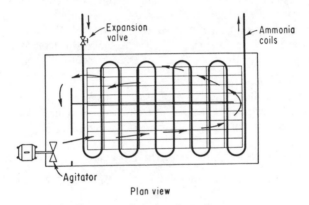

Plan view

Fig. 15-5. Brine around coils removes heat from water in cans.

utilizing large brine tanks with groups of tapered rectangular cans suspended in the cold brine. The brine is cooled by submerged pipe coils or is sometimes pumped through a shell-and-tube brine cooler. Circulation of the brine around the cans is maintained by motor-driven agitators. After a period of 24 to 48 hr the water in the cans is frozen, and the cans are removed by a crane and transported to a dip tank at one end of the brine tank. Here the cans are submerged for a few minutes in warm water until the ice is thawed loose. They are then up-ended, allowing the ice to slide out. The cans are then refilled with fresh water and replaced in the brine tank.

Q   What is the difference between white ice and clear ice?
A   White ice is ice that has formed without some form of agitation. The white or milky appearance is due to the dissolved air and gases and other impurities in the water that form millions of tiny bubbles as the water freezes. To make clear ice, these bubbles must be swept away from

the ice surface as rapidly as they are formed. This is usually accomplished by introducing a continuous stream of compressed air into the bottom of the can through a small tube provided for this purpose. The air bubbling through the water provides sufficient agitation to sweep the surface of the ice as it freezes, removing the small bubbles. When nearly all the water in the can is frozen, the core, or center portion containing the concentrated solids, is pumped out and replaced with clean water.

Q   What are some of the other methods of making ice, and where are they used?

A   Several systems for the manufacture of ice in snow, flake, cube, or briquette form have been developed. They have the following advantages: they produce ice in the form in which it is to be used; they are available in small sizes; they are automatic in operation; they produce ice within minutes of starting; and they require low capital investment in comparison with their capacity. These systems find extensive use in dairy plants, meat plants, and chemical processing plants. The smaller sizes of ice makers find wide use in restaurants and hotels.

Q   In what other field is artificial ice replacing natural ice?

A   Artificial ice is replacing natural ice in the field of sports. Within the past few years artificial ice-skating rinks have made the pleasures of ice skating available to thousands of people in many areas on a year-round basis. Refrigeration capacity is usually provided at the rate of 1 ton of refrigeration per 100 sq ft of skating surface. One-and-a-quarter-inch pipe is laid on 4- or 4½-inch centers with cold brine or cold refrigerant pumped through the pipe coils. Pipe coils may be imbedded in concrete or terrazzo or merely covered with sand.

Q   What is dry ice?

A   Dry ice is the solidified or frozen form of carbon dioxide. Its temperature is $-109.3°F$ and it absorbs 274 Btu per pound at $32°F$, giving it nearly twice the heat-absorbing capacity of water ice. It finds wide use in the storage and shipment of frozen foods and ice cream.

Q   How is mechanical refrigeration used in engineering construction?

A   Artificial cooling is used in the construction of large concrete structures, such as dams, both to cool the components of the concrete before and during mixing and to cool the water to be circulated through the mass to speed cooling by absorbing the heat of hydration. Refrigeration is also used for sinking shafts through water-bearing strata or quicksand where pumping is impractical. In this method pipes are sunk into the soil to the required depth, and cold brine is circulated until a cofferdam of frozen earth is built up. The center can then be excavated with safety.

Q What are some of the basic uses for mechanical refrigeration in the chemical industry?

A Modern chemical processes use refrigeration in the following ways: to control the rate of reaction of chemicals; to control the rate of fermentation, of liquefaction of gases and vapors, and of solidification of liquids; to absorb the heat of reactions; to prevent loss of product; and many more. The field of refrigeration is large and varied, requiring temperatures from +50°F to near absolute zero.

SUGGESTED READING

Reprints sold by *Power* magazine, costing $1.20 or less
   Refrigeration, 20 pages
   Power Handbook, 64 pages

Books
   Elonka, Steve: *Plant Operators' Manual*, rev. ed., McGraw-Hill Book Company, New York, 1965.
   Elonka, Steve, and Joseph F. Robinson: *Standard Plant Operator's Questions and Answers*, vols. I and II, McGraw-Hill Book Company, New York, 1959.
   Elonka, Steve, and Alonzo R. Parsons: *Standard Instrumentation Questions and Answers*, vols. I and II, McGraw-Hill Book Company, New York, 1962.
   ASHRAE *Applications Data Book*, American Society of Heating, Refrigerating and Air-Conditioning Engineers, 1972.

16

COOLING TOWERS

Q How is heat removed and dissipated from the compressed refrigerant?
A Heat is usually removed from the compressed refrigerant by transferring it to water in a heat exchanger where it is dissipated in several ways. If the system is near a river or lake, cooling water is taken from and returned to this natural body of water. If a ground well is used, water is circulated from the well and back to a disposal well in the ground. If city water is used, it is often discharged into the sewer; but wasting this heated water is very costly and in many localities illegal.

Q Name several reasons for reusing cooling water.
A The most important reason for reusing cooling water is that few plants are lucky enough to be near a limitless water supply. Another is that using city water for cooling is costly unless it is reused. Also, most water contains dissolved salts; using a continuous supply of such raw water quickly coats the heat exchanger with scale.

Q How does a cooling tower dissipate heat?
A The water in a cooling tower is cooled by exposure to air after each passage or cycle through the condenser, allowing the same water to be reused a number of times.

Q What is atmospheric water cooling?
A In this method the waste-heat load in the cooling water is transferred to the atmosphere. This is done by bringing the water and air together indirectly (noncontact) as is done in an automobile radiator. Another method uses evaporative-cooling equipment, such as atmospheric spray ponds or cooling towers. When water is cooled by the evaporative method, about 970 Btu is lost for each pound of water evaporated. The heat taken away in the water vapor is called the latent heat of evaporation. As the air removes this vapor, it can cool the water below the atmospheric dry-bulb temperature.

Q Why is transferring heat from water to air important?
A Because water cooled by evaporation can be reused at various temperatures. Also, being cooled to below atmospheric temperature, a small amount of water can cool a much greater heat load.

Q What is cooling by evaporation?

A The best-known way is the evaporation of perspiration on the human body, which holds its normal $98\frac{1}{2}$°F temperature even in the hottest surroundings. In a cooling tower this latent heat of evaporation is the primary cooling effect produced by blowing air over wet surfaces or through sheets of falling water.

Q What is meant by wet- and dry-bulb temperatures?

A The temperature of air on a common thermometer is known as the dry-bulb temperature. The wet-bulb temperature of the same air is identical to the dry-bulb only if the relative humidity is 100 per cent. Otherwise the wet-bulb temperature is lower, as the wet-bulb instrument has a tiny water well and wick for determining relative humidity. Thus the temperature recorded by the wet-bulb instrument is lower because of chilling by evaporation, and the degree of temperature depression is translated into relative humidity. Any psychrometric chart indicates the two temperatures.

Q How do wet- and dry-bulb temperatures compare?

A At 100 per cent humidity the wet-bulb temperature is the same as the dry-bulb temperature. However, in lower humidity some of the water will evaporate, cooling the bulb and lowering its temperature below that of the dry bulb. The difference between wet- and dry-bulb temperature increases as the air gets drier.

Q Which is better for cooling towers, cool or warm air?

A Cool air is better than warm air. It is important to have a low wet-bulb temperature, which indicates (1) very cool air, (2) very low humidity, or (3) a combination of the two.

Q Can a cooling tower cool water below the wet-bulb temperature of the inlet air?

A No. The final temperature of the cooling-tower water will always be a few degrees above the wet-bulb temperature of the inlet air.

Q How can the cooling effect be increased in a cooling tower?

A The splashing droplets wet the wood laths in large cooling towers, exposing a larger area of the water to the upmoving air. The cooling effect can be speeded by (1) increasing the air velocity over the wet surfaces, (2) increasing the area of exposed wet surface, (3) lowering the barometric pressure, (4) increasing the water temperature to the tower, and (5) reducing the humidity of the air.

Q What properties of air must be considered when buying a cooling tower?

A There are seven: (1) wet-bulb temperature, (2) dry-bulb temperature, (3) humidity, (4) total heat in Btu, (5) pressure, (6) weight, and (7) velocity.

Q What wet-bulb temperature is assumed for a given locality?

A The wet-bulb temperature assumed when designing evaporative water-cooling equipment is usually close to the average maximum wet-bulb temperature for the summer months in the given location. The exact wet-bulb temperature selected depends on operating conditions, temperature limitations, and geographical location.

Q What is meant by the cooling range?

A The cooling range is the degrees Fahrenheit that the water is cooled by the water-cooling equipment (see Fig. 16-1). It is the difference in temperature between the hot water entering the cooling tower and the cooled water leaving the tower.

Q What is meant by the term "approach"?

A The approach is the difference in degrees Fahrenheit between the temperature of the cold water leaving the cooling tower and the wet-bulb temperature of the surrounding air (Fig. 16-1).

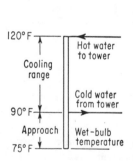

Fig. 16-1. Cooling range and approach expand schematically.

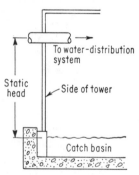

Fig. 16-2. Pumping head is static head plus velocity head and friction loss.

Q What is meant by the term "heat load"?

A The heat load is the amount of heat thrown away by the cooling tower in Btu per hour (or per minute). It is equal to the number of pounds of water circulated times the cooling range.

Q What is meant by the pumping head?

A The pumping head is the pressure needed to lift the water from its surface in the basin to the top of the tower and force it through the water-distribution system (Fig. 16-2). The pumping head is equal to the static head plus the friction loss in the distribution system and the velocity head (head needed to maintain the velocity of the water).

Q What is meant by the term "drift"?

A Drift is the small amount of unevaporated water lost from atmospheric water-cooling equipment in the form of mist or fine droplets. Drift is water entrained by the circulating air. This is the water loss independent of the water lost by evaporation. Drift loss, unlike evaporation loss, can be reduced by good design.

Q What is meant by blowdown?

A Blowdown is the continuous or intermittent wasting of a small fraction of the circulating water to prevent a concentration of chemicals in the water. The purpose of blowdown in water-cooling equipment is to reduce the solids, or hardness. This reduces the scale-forming tendencies of the water.

Q What is meant by make-up?

A Make-up is the water required to replace the water that is lost by evaporation, drift, blowdown, and small leaks.

Q Describe a cooling pond and some of its advantages.

A Water is economically cooled by circulating it through a cooling pond if the pond has enough surface exposed to the atmosphere. Ponds are most effective when only a little water is to be cooled. It is helpful if the land is cheap nearby or if there is a natural pond. The surface should receive warm water at one end and deliver cooled water from near the bottom at the other end. The cooling rate depends upon the surface area of the pond, the difference in temperature between the water and air, the velocity and humidity of the air, and the length of the air path per unit area of pond surface. The depth of the pond should be from 2 to 4 ft, and the volume should be enough to meet load variations.

Q How does a spray pond work?

A See Fig. 16-3. A spray pond increases the surface area of a cooling pond by spraying water into the atmosphere from nozzles. Tiny water particles are brought into close contact with the air. This greatly increases the surface exposed per unit weight cooled. The evaporation and resulting cooling are fast. For a given duty the spray pond needs only from 1 to 10 per cent as much surface as the simple cooling pond. The nozzles operate at from 3 to 15 psi They are spaced at 8- to 15-ft intervals on rows of pipe from 15 to 20 ft apart. There must be enough pond surface to catch spray drift in high winds. But up to 6 per cent of the water cooled is lost by evaporation and drift, even with louvers.

Q What is a cooling tower?

A A cooling tower is a wind-braced housing or shell (wood, concrete, brick, or metal) with filling on the inside. The water to be cooled is pumped into a distributing system at the top of the tower from which it drops in thin sheets or sprays to the filling. The filling is arranged so

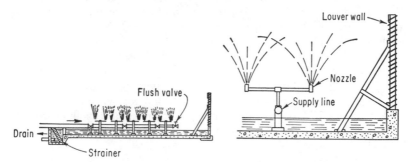

Fig. 16-3. Spray pond (above) is on ground, but can also be on roof. Sketch shows detail.

that the water spreads out to expose new surfaces to air flowing through the towers. The cooled water is collected in a reservoir or sump. The air passing through the tower becomes partly or completely saturated by evaporation from a portion of the water. This evaporation is mostly what cools the water. Cooling is usually from 4 to 10°F of wet-bulb temperature.

Q What material is the filling for cooling towers made of?

A Filling is made of boards or lath, hollow tiles, metal sheets, asbestos-cement, plastic, or special grids. Redwood, cypress, and other woods that don't deteriorate rapidly in water are used in some areas.

Q Upon what does the area of the filling's evaporating surface depend?

A The filling's exposed evaporating surface, per unit volume of tower, depends on the type of tower and how the filling is arranged. There is usually from 5 to 20 sq ft per cu ft of space filled. The area free for air flow is from 65 to 85 per cent of the total cross-sectional area occupied by the filling. The velocity of the air flow through the tower depends on the operating conditions. It is usually from 100 to 700 fpm.

Q How are cooling towers classified?

A Cooling towers are usually classified by the way their housing is arranged and by their draft-producing methods. They are broadly classified as (1) natural draft, (2) induced draft, and (3) forced draft. Natural-draft towers may be further classified as spray towers and deck towers.

Q Describe an atmospheric cooling tower.
A See Fig. 16-4. The atmospheric spray-filled cooling tower (or open natural-draft tower) is the simplest cooling-tower design. Air enters through louvered sides, which prevent water from blowing out, and flows through in a transverse direction. The air circulation depends on the velocity of the wind. With 5 mph wind velocity these towers can cool up to 1.5 gpm of water per square foot of active horizontal area. They are used for water-cooling needs of less than 30,000 Btu per min. Ranging from 6 to 15 ft high, these towers have a narrow spray pond with nozzles above it and a high louver fence. The nozzles spray downward. The louvers are always wet; hence they add to the water surface exposed to the cooling air. These towers are used where (1) equipment served can stand a few degrees rise in cold-water temperature at low or zero wind velocities, (2) drift from cooling tower isn't objectionable, and (3) tower can be placed so that the wind isn't cut off by other buildings, trees, etc.

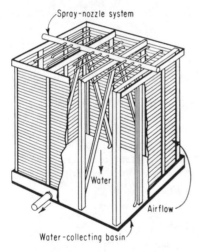

Fig. 16-4. Atmospheric cooling tower.

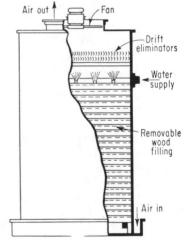

Fig. 16-5. Mechanical-draft tower.

Q Describe the mechanical-draft cooling tower.
A See Fig. 16-5. A mechanical-draft cooling tower is a vertical shell of wood, metal, Transite, or masonry. Water is distributed near the top of the shell and falls to a collecting basin. As it falls, it passes through air that is circulated from the bottom to the top of the shell by either forced- or induced-draft fans. Because it passes against the flow of water, a given quantity of air picks up more heat than does an equal quantity of air on natural-draft equipment. So less air is needed to cool the same

amount of water. Since the air is supplied by fans, the air quantity must be held to a minimum for low operating cost. To prevent excessive water loss and nuisance caused by drift, baffles are usually installed at the top of the tower. For a given ground area mechanical-draft towers may have from two to five times the cooling capacity of chimney towers. The velocity of air flow through the tower is usually between 300 and 700 fpm. Loss of water by evaporation and drift is very small.

The efficiency of these towers is improved by increasing the wood filling, the height or area of the tower, or the quantity of air supplied. Increasing the tower's height increases the time that the air is in contact with the water without increasing the need for more fan power. And, increasing the tower's area while keeping the fan power constant increases the quantity of air and air-water contact time because of lower velocity. The water-surface area in contact with the air is increased in both cases. Increasing the quantity of air decreases the time that the air is in contact with the water. But since more air passes through the tower, the average difference between the water temperature and the wet-bulb temperature of the air is increased. This speeds up the heat-transfer rate. More air means an increase in fan power. Since these towers don't depend on wind velocity, it is possible to design them for more exacting performance. They require less space and less piping than do atmospheric cooling towers. Pumping heads vary from 11 to 26 ft. The over-all plant economy from the lower water temperature supplied by these units usually offsets their added operating expense and high initial cost.

Q Describe a forced-draft cooling tower.

A See Fig. 16-6. A forced-draft cooling tower works well with corrosive

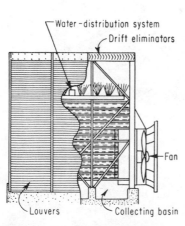

Fig. 16-6. Forced-draft cooling tower.

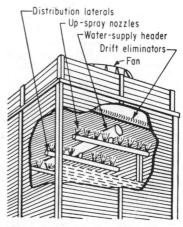

Fig. 16-7. Counterflow induced-draft cooling tower.

waters; since the fan can be near the ground, the parts most likely to corrode are easily serviced. However, the fan maintenance and depreciation costs of these units are high. The main objective in these units is to have heated air leave the top of the tower at a low velocity. But at times the air recirculates to the inlet of the fan, and with unfavorable winds this cuts efficiency as much as 20 per cent. During cold weather recirculation can cause ice formations on nearby equipment, buildings, or in the fan ring of the tower, sometimes breaking the fan. Because the fan size on these units is limited to a 12 ft diameter or less, more fans, motors, starters, and wiring are needed for larger towers than for the same size of induced-draft towers. The latter can use fans with up to 60 ft diameter. However, forced-draft cooling towers offer a neater appearance, are more adaptable to architectural surroundings, and, by locating the fan at the top of the tower, noise is minimized.

Q   Describe a counterflow induced-draft cooling tower.

A   See Fig. 16-7. In a counterflow induced-draft cooling tower the fan is mounted at the top. The air movement is vertical, across the filling, and upward at a high velocity to prevent recirculation. If the unit is to handle small loads, the fan is mounted at one end to give a horizontal air flow.

Q   Describe a double-flow cooling tower.

A   See Fig. 16-8. Demands for a compact design, better construction,

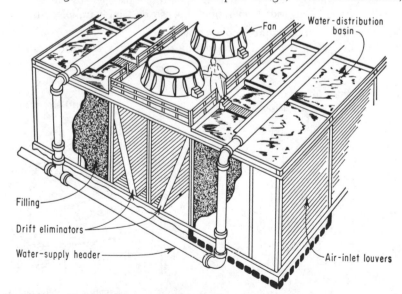

Fan

Water-distribution basin

Filling

Drift eliminators

Water-supply header

Air-inlet louvers

Fig. 16-8. Double-flow cooling tower packed with a filling.

lower cost, large capacity, more flexible operation, and improved all-round performance produced the double-flow cooling tower. It is also called a common-flow tower. The air flow is horizontal, and fans are centered along the top. Each fan draws air through two cells that are paired to a suction chamber partitioned midway beneath the fan and fitted with drift eliminators to turn the air upward toward the fan outlet. Double-flow towers use low pumping heads, which vary from 11 to 26 ft.

The operating advantages of double-flow cooling towers are: (1) Horizontal (crossflow) air movement. As the water falls in a cascade of small drops over the filling and across the air stream, there is less resistance to air flow and therefore a lower draft loss. (2) Air travel is longer than in a conventional design. (3) The open water-distribution basin is accessible for cleaning during operation. (4) The closely spaced wood and diffusion deck under the water basin result in uniform water distribution to the wood filling. (5) Water loading in most cooling towers has a maximum of 6 gpm per sq ft caused by a blanketing spray effect. But a heavier loading, up to 10 gpm per sq ft, is possible in double-flow towers for steam-condensing service. (6) Modern double-flow units occupy less than one-twentieth of the area needed by a spray pond for equivalent service.

Q Sketch a typical cooling-tower system.
A See Fig. 16-9. Here, water from the sump under the tower flows to

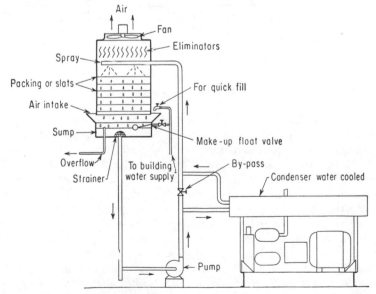

Fig. 16-9. Typical cooling-tower system.

the circulating pump by gravity. The pump forces the water through the condenser where it picks up heat from the condensing refrigerant. Warm water from the condenser goes to the sprays in the cooling tower where it is broken into drops and falls over the wood slats or packing. The drops, while traveling through the air and over the tower packing, are exposed to a blast of air induced by the cooling-tower fan. Although part of the water evaporates in this air stream, most of it is cooled. This cooled water collects in the sump and flows to the pump for reuse.

Q   Describe evaporative condensers, and explain why they are often used instead of cooling towers.

A   See Fig. 16-10. Evaporative condensers overcome some of the serious disadvantages of cooling towers. As the sketch shows, the water-cooled

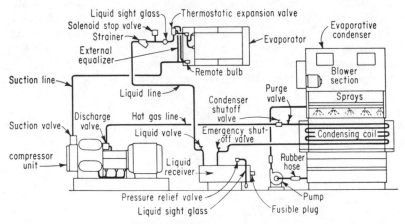

Fig. 16-10. Evaporative-condenser installation is modern, compact.

condenser is missing. Hot gas from the discharge of the compressor goes directly to the condensing coil inside the evaporative condenser. There, it condenses to a liquid and runs off to the receiver. Heat is removed from the condensing gas by water evaporating directly on the coil. The evaporation is speeded by an air stream induced by the fans. Then the water remaining after the evaporation drips to the sump, collects, and flows to the spray pump for recirculation. The pump horsepower of an evaporative condenser is much less than that for a cooling tower. One reason for this is that there is shorter piping and less spray water to resist water flow. Also, water losses are much less than the water losses from a natural-draft tower. Because this condenser creates its own air movement, it can be installed either outside or indoors. But if installed indoors, it must have a fresh-air supply and a way to discharge humid air.

Q What are some precautions to take when installing cooling towers indoors?

A See Fig. 16-11. There must be an abundant supply of fresh air reaching the tower. Air must enter the tower through fixed open windows, in-

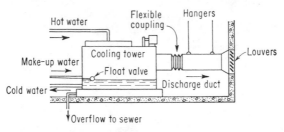

Fig. 16-11. Indoor cooling tower installation.

take ducts, or other openings, such as driveways going into basements of large buildings. The free air area of the opening should be as large as the air intake of the tower. If the tower intake duct is used, it should be of the same continuous size as the air intake area of the cooling tower. Hot-air discharge from a cooling tower must be ducted to the outdoors. The duct should be pitched toward the tower and should be well sealed and as short as possible. Area changes, joints, and elbows should be kept to a minimum. The duct should be self-supporting and equipped with an access door for easy cleaning. The duct draft loss should not exceed 0 10 in. of water pressure at design air-flow rate. A flexible coupling as in the duct shown is most important.

Q Describe a hyperbolic cooling tower.

A See Fig. 16-12. These are large chimney-draft or closed natural-draft towers up to 340 ft high and 260 ft in diameter at the base. They are usually used by large power plants of 100 megawatts and over. The chimney, or shell, is commonly of reinforced concrete and hyperbolic in shape. The upper half of the shell has a throat, or smaller diameter. The shell is supported on legs about 20 ft high. Inside the chimney is a cooling stack of wood over which the warm water is distributed to present the maximum surface area to the air. The stack is of such depth that the water flowing through has enough time for the desired cooling. The chimney around the cooling stack has openings at the base high enough to give the required draft and with enough area to give the desired rate of air flow. The base has a pond to collect the cooling water. The base also has water suction and delivery pipework or channels and the needed distribution penstocks. Atmospheric air flows in from outside around the base and up inside to the top, induced by the chimney action.

Q   What are some of the outstanding advantages of hyperbolic cooling towers?

A   (1) Because of their height, ground fogging and recirculation of warm, moist vapor do not occur. (2) Atmospheric air for direct water

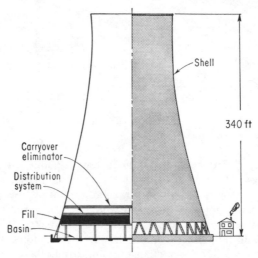

Fig. 16-12. Hyperbolic cooling towers are large chimney-draft structures.

cooling is induced through the tower by thermal power created by heating and humidifying the air. (3) Operation is assured even at zero wind conditions, which is contrary to the blow-through atmospheric tower. (4) No moving equipment is needed to create air flow. Power is used only for pumping. (5) Inspection and maintenance are cut to a minimum.

In general, the hyperbolic natural-draft tower is at its best when the difference between the desired cold-water temperature and the wet-bulb temperature is equal to or greater than the difference between the hot-water temperature and the cold-water temperature. In cooling-tower language this would be an approach equal to or greater than the range. Air flow through a natural-draft tower increases during the cooler months. This increase will offset the loss of heat transfer driving force at the lower wet-bulb temperature. This loss cannot be offset in mechanical-draft towers without increasing the air rate at the expense of using more fan horsepower.

Q   Describe the operation of a hyperbolic cooling tower.

A   These towers usually operate in parallel. The discharge pipe from the condenser delivers water to an upper culvert, which opens into the annular culverts around each tower. The amount of water diverted to any one tower can be controlled by penstocks. A simple sluice valve controls the intake to each distributor pipe within the cooling stack. After coming through the tower, the cooled water collects in a pond underneath. The walls of the pond are about 10 ft high. A wall usually divides the pond in halves, so one half can be shut off and cleaned while the other half works. All the ponds connect through penstocks to the suction culvert, and both the suction and discharge culverts can be sectionalized. Any tower can be completely isolated from its neighbor. A drain culvert usually runs along the rear of each pond for bleeding water to reduce concentration from evaporation. Penstocks control the drains.

Q What are some causes of spotty performance in a new cooling tower?
A Only a few things might cause eccentric performance in a new tower. Check for concentration of air or water flow, blanketing by water, too much wind, or irregular load conditions.

Q What can upset cooling-tower performance?
A For best performance a tower needs the right quantity of air at the proper temperature, the right amount of water under the original design temperatures, and atmospheric conditions as outlined in the specifications. If the water pumps are running at the right speed, the water supply is right, and the fans are turning at proper rpm, look for fouling. Fouling can be from algae or mud deposits. Then the cooling effect drops to a fraction of the original. Nozzles might be plugged. Foreign matter build-up on the tower surfaces can upset the air flow. Modern towers permit the removal of a section of fill through openings in the base. Dirty towers can't operate at design specifications.

Q Explain how to prevent ice from forming in induced-draft towers.
A Run two-speed motors at half speed to retard ice formation. If the unit is a multicell design, shut down some fans. Cover the upper portion of the louvered area to limit the air flow. Reduce the tower's water flow and stop one or more fans of the multicell unit. Bypass water to part of the cooling tower, and stop one or more fans.

Q Explain how to prevent ice from forming on louvers and wood filling.
A Reverse the fan motor (for limited time only) to blow warm air out the louvers. But be careful, because some speed reducers don't supply oil to the upper bearing when reversed. Stop some fans temporarily but don't stop the water. When the cells thaw out, start the fans and repeat the process in other cells. Cover the upper portions of the louvered area to limit the air flow.

Q Fogging can be a problem with cooling towers. How is it caused?
A Fogging occurs under certain atmospheric conditions. By definition, the dew point of the air is the temperature at which air reaches a state of saturation when cooled. When air is cooled to its dew point, moisture begins condensing and fog results. Fog is most objectionable to buildings, highways, etc. When a tower ices up, the fans move less air, so air leaves the tower at a low velocity and a high temperature. Much fog is caused when the warm tower air mixes with cold outside air. Fog is airborne droplets formed by vapor condensation. As warm and nearly saturated air leaves the cooling tower in cold weather, it mixes with the surrounding air and then cools, condensing to vapor and forming fog. Fog is often worse during mild winter weather of 50 to 60°F than with subzero temperatures.

Q Explain what effect ice has on cooling-tower operation.
A While extremely cold water usually doesn't increase tower performance much, it does increase operating hazards. Water-cooling towers operating in freezing weather are subject to ice formation at air inlets. Small amounts of water vapor are likely to freeze on the inlet louvers and nearby wood filling. Ice starts to form on the lower louvers and climbs upward. This restricts the inlet area, reduces the air flow, and increases the temperature of the water circulated through the tower. Most of the circulating water is seldom cooled to freezing temperature. Actually, the cold-water temperature seldom gets lower than 60°F, except in towers used for special operating conditions. For this reason ice forms only on the parts of the tower that are lightly wetted by fine drops which splash toward the entering air stream.

Q Explain how to operate a cooling tower in winter to prevent ice formation.
A To prevent icing during cold weather, keep the temperature of the circulating water as high as practicable. A temperature-control valve may be used to bypass some hot water to the cold-water basin. Usually, ice formation in water-cooling towers can be prevented, controlled, and removed by varying the tower air flow. When the unit is operated off and on during winter weather, drain the water from all exposed piping and basins. This ensures protection against freezing and corrosion. Leave the basin drains open during winter shutdown to allow rain and melted snow to escape.

Q How long is it safe to run a cooling-tower fan in reverse?
A Don't run a fan in reverse for more than 5 to 10 min. Few speed reducers supply oil to the upper fan-shaft bearing when reversed. The lube pump may not deliver oil when reversed, which would cause bearing

failure. Thrust bearings and screw threads can also cause major troubles when the fan is reversed. Keep this in mind and proceed with caution unless you are sure of the proper operation.

Q What mechanical maintenance does a cooling tower need?
A Set up a preventive maintenance schedule for your cooling tower. Grease and oil the bearings and check the motor insulation yearly. Refill the speed reducers every 3,000 hr. Look for oil leaks around the fan driveshaft caused by worn oil seals. Keep the fan blades and hubs painted to avoid corrosion. Tighten the blade clamps and hub bolts and rebalance the fans when necessary. Check the clearances, lubrication, and air gaps periodically.

Q What structural maintenance does a cooling tower require?
A Replace or repair damaged cooling-tower parts immediately. Remove scale, dirt, bugs, and debris; then clean and paint the metal parts. Tighten any loose bolts, but remember that the wood swells. Drift eliminators reduce the air flow when they are dirty, so keep them clean. Wash the dirt from the wood filling slats.

Q How do you shut down a cooling tower?
A When shutting down a cooling tower, drain it to prevent freezing and corrosion. Leave the drain open so that rain and melted snow can escape. Once a week run the fans for 5 min in order to keep the upper fan-shaft bearing oiled. Protect the metal parts from corrosion. While the tower is shut down, do any necessary maintenance and repair work so that it will be ready for the next season.

Q What are some of the limitations of cooling towers?
A Water that is cooled by spraying into the air cannot be cooled below the wet-bulb temperature of the air. Therefore, the condensing temperature at which the system will operate must be kept slightly above the wet-bulb temperature of the air. High-efficiency forced- or induced-draft cooling towers will cool the condenser water to a point within 5 to 8°F of the prevailing wet-bulb temperature. Natural-draft towers seldom approach closer than 10 or 12°F of the wet-bulb temperature of the air. So when selecting a cooling tower, it is best to increase the usual "design" wet-bulb temperature by 5 per cent.

Q What are some disadvantages of cooling towers when compared with evaporative condensers?
A Natural-draft cooling towers depend upon the natural movement of the air, which is often unreliable. The adjacent structures or natural obstacles may deflect the breeze from the tower. Or, a stiff breeze may

cause an unusually large amount of water to be carried away from the natural-draft tower in the form of drift or air-entrained spray. The draft is often a nuisance to adjoining property. Because forced- or induced-draft towers depend on fans, they avoid these troubles. But in freezing weather the air must be reduced to a point where the spray water won't freeze in the sump, on the packing, or in the discharge of the tower. Furthermore, you must protect the overflow and make-up-water piping. In severe weather you may need an auxiliary tank in the heated part of the building to accumulate the reserve water instead of using the sump in the tower. This extra tank allows the spray water to drain from the tower immediately.

Q What six steps are needed in sizing the pump and piping of any cooling tower installation?
A (1) Determine the amount of water in gallons per minute to be circulated to the tower; (2) make a layout of the complete piping system; (3) determine the static head and pressure drop through all units other than the pipe and fittings; (4) select the pump; (5) size the pipe; and (6) check the sizes of the pump and the pipe.

Q How do you determine the amount of water in gallons per minute to be circulated through a cooling tower?
A For air-conditioning and refrigeration cooling towers, the water flow required can usually be determined from the tower manufacturer's data. The average cooling-tower application requires 3 to 5 gpm per ton. For cooling jobs other than air conditioning and refrigeration the amount of water needed must be determined from the cooling load.

Q Do cooling towers need automatic sprinklers for fire protection?
A If induced-draft cooling towers are made of wood, have a roof over their fans, and have their supporting structure on the roof, they need automatic sprinklers. But if towers are the forced-draft type with no roof, and if the fans are at the base to force the air up through the tower, they do not need automatic sprinklers.

Q Name six possible ways to increase cooling-tower capacity.
A When space is not available for a second cooling tower, there are six possible ways to increase the capacity of the present tower: (1) Fill of a newer type often increases the fill area, as well as the time the water's surface is exposed to upcoming air. Up to 20 per cent additional capacity may be gained. (2) A distribution system, the positive-pressure spray type for example, remains balanced as the tower ages. But the spray patterns may be improved by up to 15 per cent of capacity. (3) Drift eliminators and inlet louvers in some older designs restrict air flow.

Blades staggered at a 60° angle are better, and may give up to 5 per cent better performance. (4) Mechanical equipment moves air through a tower; thus fan blade angle, speed, and size of motor are all variables. Improving these may raise capacity by 10 per cent. (5) The fan stack may raise capacity without changing the motor size. A proper velocity recovery-type stack can delivery up to 7 per cent more air. (6) Partitioning is very important in large multicell towers because it prevents short-circuiting air if one fan does little work. All in all, a 20 per cent increase in capacity can be achieved in some towers by making these six changes.

Q How are cooling towers protected from freezing during winter operation?

A Today we cool buildings and run air-conditioning equipment during freezing weather. The expanding demand for cooling on a 12-month basis means that cooling towers have to operate in freezing as well as in warm weather. In some large buildings, all-year cooling is required because the inner core areas must dissipate substantial amounts of heat generated by office machines, high-intensity lighting, and high-occupancy loads. Changeable weather may call for heating for a few days, followed by several warm days which require cooling. Electric heat is often used to protect cooling towers against freezing. Special electric heaters and thermostatic and low-water controls are used, as shown in Fig. 16-13.

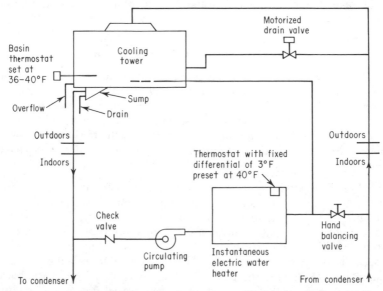

Fig. 16-13. Power wiring for the tower fan motor is often large enough to handle immersion-type electric heaters (not simultaneously).

SPRING STARTUP OF COOLING TOWERS

Q Explain how to prepare cooling towers for summer operation.

A Before filling the system with cooling water, clean out and hose down the collecting pan and all accessible interior surfaces of the tower to remove airborne grime that has accumulated during the winter. (Cooling towers in city locations often require additional flushing during the operating season.) Adjust the float valve (or ball cock) so that it closes completely when the water level in the collecting pan is about 1 in. below the overflow point. Keep the ball cock tightly closed when sprays are off. Make sure there is no overflow when sprays are operating. Also, keep the water level low enough so the pan won't overflow when the tower is shut down and water drains from the upper parts.

Remember, ball-cock discharge should always be *above* the water level in the collecting pan because a submerged inlet is an *illegal* (and dangerous) cross-connection with the drinking-water system. Adjust circulating-water pump packing gland to permit just enough leakage (if jam-type packing is used) for lubrication. Check the manufacturer's recommendations and replace the packing if necessary.

Inspect the perforated head pans, spray nozzles, and strainers. Be sure to clean them if necessary. (Of course, the best time to do this is after fall shutdown.) Then check the refrigerant condensers and, if possible, take off their heads and clean all sediment and scale from the water side.

SUMMER OPERATION MAINTENANCE

Q What attention do cooling towers need during summer operation?

A Cooling water scrubs dirt, soot, fibers, fly ash, and other solid particles from the air. These solids not only accumulate in the collecting pan, but work past strainers and clog pipes and condenser tubes. A dirty cooling-water system favors slime and algae growth. If you don't want to shut down for chemical cleaning, you must do periodic mechanical cleaning.

Q What systematic checks are necessary for cooling-tower operation?

A There are seven: (1) clean spray nozzles and head pans to prevent buildup; (2) clean strainers and screens when necessary; (3) adjust pump packing to permit only enough leakage for lubrication; (4) check for a vortex at the suction of the pump when the cooling tower is operating— adjust the water level or install baffles to keep air from being drawn into the piping; (5) adjust the ball cock if necessary; (6) inspect the bottom of the collecting pan, drain, and flush accumulated dirt; and (7) inspect the interior of the equipment. Hose down to remove loose deposits on cooling-tower fill or eliminators.

Q What encourages algae growth in cooling towers?

A Because algae are colored, they need sunlight to grow. An opaque cover over the tower head pan to keep out sunlight often prevents growth. Some cooling-tower manufacturers supply covers. As soon as slime or algae is noticed, call your water-treatment company.

Detergent cleaning may restore the cooling tower to full operation within hours. But cleaning must be done with detergent, not acid. Acid cleaning is needed to remove carbonate scale, but isn't usually effective for dirt or slime.

Q What attention do electric motors used for cooling towers need?

A Some pump motors have factory-lubricated bearings and need no attention. If the motor has grease cups or fittings, lube the bearings with a soda-base grease. Examine the bearings for wear. Remove them every two years and clean them with a mineral spirit, then regrease.

Check the oil level, and if oil is needed, add a good grade of SAE 20 compressor oil.

NOTE: Do not use automobile crankcase oil.

For winter operation below zero, it is very important to use an oil that will pour at a temperature at least 20°F below the operating temperature. Through a grease fitting, add a special waterproof grease to the grease compartment of the water-pump seal. Lubricate fan bearings once a year; disassemble and clean them every two years. Be sure to wipe grease or oil from fan bearings and belts.

LAYING UP COOLING TOWERS FOR WINTER

Q Explain how to lay up a cooling tower after the cooling season is over.

A (1) Take the float valves and ball cocks apart for cleaning. Adjust them so that they close completely when the water level in the pan or basin is about 1 in. below the overflow point. (2) Drain and flush pans, pipelines, pumps, and heat-exchanger equipment. Clean perforated head pans and spray nozzles. For best results, seal all piping openings to prevent corrosion. (3) Clean strainers, screens, and pump strainer baskets, then remove overflow pipe. (4) Clean the water side of the refrigerant condenser. If heads can't be taken off, back-flush to waste with high-velocity water. After checking tube sheets and water boxes, recoat them if necessary. (5) Screen louvers and fan openings must be covered. (6) Coat metal parts exposed to weather with alkaliproof paint.

NOTE: Don't paint spray nozzles or heat-transfer surfaces.

(7) If the system is closed, remove the strainers and clean. If it is laid up full or partly filled, take precautions against freezing. Drain water

coils which will be exposed to freezing temperatures, then flush them with antifreeze solution. If units have low-temperature alarms or antifreeze shutoffs, check and calibrate such controls before winter. (8) On gear-driven units, operate the gear and fan for a few minutes each week during shutdown. This will keep gears and bearings coated with oil and prevent rusting.

NOTE: Don't circulate water during this operation.

SUGGESTED READING

Reprints sold by *Power* magazine, costing $1.20 or less
   Handbook on Fans, 16 pages
   Pumps, 32 pages
   Heat Exchangers, 32 pages
   Balancing Rotating Equipment, 24 pages
   Vibration Isolation, 16 pages

Books
   Elonka, Steve, and Joseph F. Robinson: *Standard Plant Operator's Questions and Answers*, vol. II, McGraw-Hill Book Company, New York, 1959.
   Elonka, Steve: *Plant Operators' Manual*, rev. ed., McGraw-Hill Book Company, New York, 1965.
   Elonka, Steve, and Alonzo R. Parsons: *Standard Instrumentation Questions and Answers*, vols. I and II, McGraw-Hill Book Company, New York, 1962.
   Elonka, Steve, and Julian L. Bernstein: *Standard Electronics Questions and Answers*, vols. I and II, McGraw-Hill Book Company, New York, 1964.

17

AIR CONDITIONING

Q What is complete air conditioning?
A Complete air conditioning means heating, cooling, humidifying and dehumidifying, circulating, and cleaning the air.

Q What factors affect the conditioning load?
A A group of complex factors affect the conditioning load. They are (1) heat transmission, (2) solar radiation or sun effect, (3) people, (4) light and power equipment, (5) ventilation air or infiltration, (6) product load, and (7) miscellaneous.

Q What is meant by heat transmission?
A Heat transmission is the heat flow through walls, floors, windows, ceilings, and roof. It comes about from a temperature difference between the inside air-conditioned space and the outside atmosphere. Heat flows in when the temperature is higher outside. This unwanted heat must be removed by cool air.

Q What should be known about outside temperatures?
A Weather conditions make up most of the heat-transmission load in conditioned spaces. Local weather bureaus forecast valuable information on this subject. An operating engineer can use forecasts to plan operations ahead. Insulating spaces or buildings against transmission loads reduces the load on air-conditioning equipment.

Q What effect does sunlight have on an air-conditioned space?
A Windows exposed to sunlight transmit most of the solar radiation. This in turn is absorbed by furniture, fixtures, and flooring. Solar radiation can be reduced with blinds, awnings, or light-colored paint on the outside of building walls. Painting roofs aluminum or spraying them with water during sun periods reduces radiation.

Q What effect does the "people load" have on the system?
A Large places, such as theaters, filling with people suddenly can add to the heat load sharply. The area to be cooled should be brought to the

proper temperature before people arrive. It is much easier to hold conditions after a precooling period than it is to bring down conditions by starting up a cooling system after people arrive.

Q   How do lights affect the cooling load?
A   Lights add very little heat to average spaces, but display lighting in department stores often makes up a large portion of the load. The lighting load can be estimated from meter readings or by multiplying the total watts by 3.4.

Q   What effect does infiltration or ventilation have on the load?
A   This represents an important load factor. It varies with weather conditions, people load, and building construction. Tightness of doors, windows, etc., are important preventive measures.

Q   What is air?
A   Pure, dry air is a mixture of oxygen and nitrogen, plus small amounts of rare gases, such as argon. The air around us also contains moisture in varying amounts. Air to be conditioned might therefore be called a mixture of air and steam.

Q   How are air temperature and humidity related?
A   Air temperature and humidity are related through the basic properties of steam. For any given air-water temperature, each cubic foot of air can hold a specific weight of water vapor: this is the saturation-temperature dew point.

Q   How is water vapor superheated?
A   When air temperature is above saturation temperature, the water vapor is superheated, making the air capable of holding a greater weight of water. This degree of water saturation is charted as per cent humidity or per cent relative humidity.

Q   What is saturation temperature?
A   For any given pressure there is one temperature—the saturation temperature—at which steam starts to vaporize or condense. Let us say that 1 cu ft of moist air at 70°F contains 0.0004 lb of moisture. But the same air can hold 0.0011 lb of steam. The reason why it is less is that the steam is superheated. In this case, $70 - 40 = 30°F$ of superheat.

Q   What is relative humidity?
A   Because the cubic foot of air in the above problem holds less moisture than it is capable of holding, we have relative humidity. Here, we have 0.0004 lb of moisture when we could have 0.0011 lb. So the ratio is 4 to 11, or 36 per cent relative humidity. This is based on volume, which is cubic feet of moist air.

Q What is dew point?

A If we take the mixture, with 0.0004 lb of moisture in a cubic foot, and cool it down to 40°F, it will hold all the moisture that 1 cu ft can hold at that temperature. Then we say it has 100 per cent relative humidity because the mixture is completely saturated. If the mixture is cooled still further, some of the steam will condense. In this case condensation will start at 40°F, which is the dew-point temperature. The dew point of any mixture of air and water vapor therefore depends on the amount of moisture present.

Q What is wet-bulb temperature?

A To take the wet-bulb temperature, you need a wet-bulb thermometer and a psychrometric chart. The thermometer has a bulb that is covered with wetted silk gauze and placed in the air stream. Some of the water in the gauze will evaporate. Because vaporizing takes heat from the remaining water, the water temperature will drop. The amount of this temperature drop depends on the dryness and temperature of the air. The thermometer shows the wet-bulb temperature and tells us the amount of moisture in the air. Then, by checking a psychrometric chart, we learn the other things we need to know. (See discussion of latent heat and sensible heat in Chap. 1.)

Q What is total heat?

A For completely dry air the sensible heat would be the total heat. For water vapor alone its total heat would be sensible heat plus latent heat. Total heat depends on the wet-bulb temperature, just as sensible heat depends on the dry-bulb temperature and latent heat depends on the dew point.

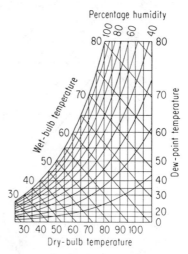

Fig. 17-1. Psychrometric chart plots moist-air conditions. (*Courtesy The Trane Co.*)

Q What is a psychrometric chart?

A See Fig. 17-1. Because moist-air mixtures always follow the same rules, a psychrometric chart has their conditions plotted. This saves time in calculations. The chart shows percentage of humidity (some charts show relative humidity), wet- and dry-bulb temperatures, and dew point. By using auxiliary scales, many useful data can be obtained from the chart.

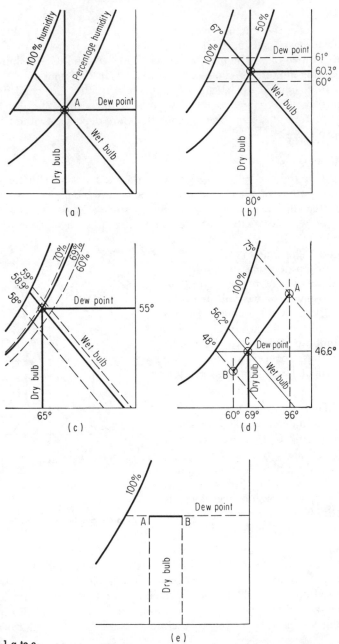

Fig. 17-1 a to e.

230

Q What does a psychrometric chart show?

A The skeleton chart (Fig. 17-1a) is a Trane Psychrometric Chart. Lines of the constant dry-bulb temperature are almost vertical. Lines of the constant dew-point temperature are horizontal. Lines of the constant wet-bulb temperature slope downward to the right. Lines of the constant percentage humidity are curved. Before using the chart, any two of the above values must be known. The remaining two can then be found on the chart.

Q How do you find humidity on a psychrometric chart?

A EXAMPLE: Given air having a dry-bulb temperature of 80°F and a wet-bulb temperature of 67°F, find the dew-point temperature and the percentage of humidity. Refer to the skeleton chart in Fig. 17-1b. Then try this problem on the psychrometric chart in Fig. 17-1. You will find that the dew-point temperature of the air is 60.3°F and that the humidity is 50 per cent.

Q How do you find the wet-bulb reading?

A EXAMPLE: Given air having a dry-bulb temperature of 65°F and a dew-point temperature of 55°F, find the wet-bulb temperature and the percentage of humidity. Refer to the skeleton chart in Fig. 17-1c. Then try this problem on the psychrometric chart. You will find that the wet-bulb temperature is 58.9°F and the humidity is 69 per cent.

Q How do you find air mixtures on the Trane chart?

A EXAMPLE: The chart can be used to determine resultant wet-bulb and dew-point temperatures of a mixture of two streams of air at different conditions. Suppose that 25 per cent of the mixture has a dry-bulb temperature of 96°F and a wet-bulb temperature of 75°F, as represented by point A in the sketch in Fig. 17-1d. The balance, or 75 per cent, has a dry-bulb temperature of 60°F and a wet-bulb temperature of 48°F, as shown by point B. The resultant dry-bulb temperature of the mixture is figured as follows:

$$0.25 \times 96 = 24°F$$
$$0.75 \times 60 = 45°F$$
$$24 + 45 = 69°F, \text{ the resultant dry-bulb temperature of the air}$$
mixture

Now, draw a straight line between points A and B. The intersection of the vertical line of 69°F dry bulb with the line from A to B gives point C. The wet-bulb and dew-point lines running through point C are the resultant wet-bulb and dew-point temperatures of the mixture. In this case the wet-bulb temperature of the mixture is 56.2°F and the dew-point temperature is 46.6°F.

Q   How do you use the Trane chart to trace air-conditioning cycles?
A   Besides giving the properties of air, the Trane chart can be used to trace air-conditioning cycles. Processes involving heating or cooling and humidification or dehumidification can be clearly followed on the chart. In tracing any process, keep this fact in mind: the dew-point temperature is constant as long as there is no change in the moisture content of the air. Thus, heating of air without changing its moisture content takes place along a horizontal line of the constant dew point. The dew-point line is determined by the initial condition of the air. In Fig. 17-1e, the air that is initially in the condition designated by point A is heated to point B along the horizontal line AB. Also, if the air is to be cooled without the moisture being condensed, the process is represented by a straight line drawn from points A and B in Fig. 17-1e.

Q   How do you find air volumes on the Trane chart?
A   See Fig. 17-2. To find the volume of 1 lb of air having any temperature and humidity, you must know the dry-bulb and dew-point temperatures of the air. The diagonal lines on the Trane chart give the volume of 1 lb of dry air. Thus, using the volume chart, air at a dry-bulb temperature of 71°F at a dew-point temperature of 59°F has a volume of 13.6 cu ft per lb.

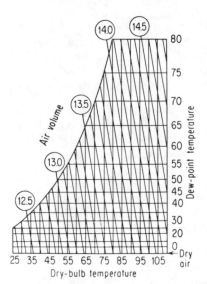

Fig. 17-2. Trane air volume chart.
(*Courtesy The Trane Co.*)

Q   Sketch a diagram of a spray-cooling unit and explain how it works.
A   See Fig. 17-3. A spray-cooling unit is the only system capable of humidifying and scrubbing air while controlling its temperature. The cycle diagram shown includes the basic parts of a spray unit. High-efficiency atomizers generate a fog bank at the air inlet. For the most effective humidifying action two or more banks of sprays are used. Since effectiveness depends upon intimate mixing of air and water vapor, the air velocity through the spray chamber usually is kept between 350 and 600 fpm and the path of air travel is long. The spray section is followed by moisture eliminators mounted in banks. These usually consist of wetted plates or scrubbers placed on close centers. A different method of

securing intimate contact between air and water uses extended wetted surfaces. Another method uses spray water distributed by capillary action through glass-fiber mats. Air passing through the thoroughly soaked mass

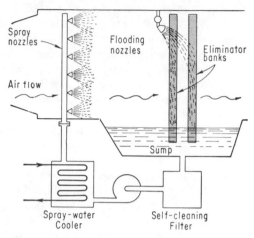

Fig. 17-3. Spray cooling works on principle that air may be cooled or dehumidified when sprayed with a suitable liquid.

of fibers comes in close contact with water over an extensive path. The result is high-efficiency humidification at the expense of higher air re-

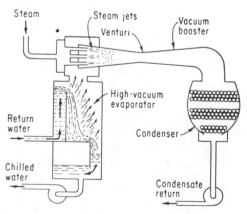

Fig. 17-4. Jet cooling system uses steam-jet ejectors and water for refrigerant.

sistance. This construction also acts like a viscous filter in cleaning passing air. Coil cooling may combine with spray equipment or handle the entire cooling-humidification job. Where centralized air cooling and extensive ductwork are not practicable, centralized refrigeration equipment can supply coolant to remote coil units.

Q Sketch a jet-cooling cycle, and explain how it works.

A See Fig. 17-4. Jet cooling uses water as the refrigerant for a coil system. The main advantages of jet cooling are cheap refrigerant and complete safety in the event of cycle leaks. Because high vacuum is needed to vaporize water, the usual apparatus uses steam-jet ejectors. Since the condenser must handle compressed water vapor plus driving steam, the cooling-water requirements are high. The system finds use where cooling water is plentiful, steam is cheap, and refrigerant temperature requirements are fairly moderate. The jet-cooling cycle uses a high-vacuum chamber to boil off part of the incoming water at a low temperature. Since there is no heat input to this part of cycle, the latent heat of vaporization—about 1050 Btu per lb in this case—is supplied by using sensible heat from the unvaporized water. The final temperature of the chilled water is restricted to 35°F or above by practical limitation on the vacuum. In the diagram of the unit shown high vacuum is created by steam-jet ejectors which pull vaporized moisture from the top of the vacuum chamber. Evaporated water and ejector steam exhaust to a normal shell-and-tube condenser. Since the condenser must absorb total energy from the relatively high-pressure ejector, the steam-condenser cooling-water needs are high. In addition, high vacuum means high steam consumption by jet ejectors. These conditions are reflected in higher operating costs.

Q Sketch an absorption unit, and explain how it works.

A See Fig. 17-5. An absorption system is based on the characteristics of liquid sorbents and uses heat for its driving force. Since a high-level heat source is not needed, the cycle finds special usefulness where low-pressure steam or waste heat is available. It is an old principle, finding increased application since the introduction of lithium bromide sorbent and the construction of large-capacity units. The units require considerable cooling-water flow and show their best relative operating efficiency under full-load conditions. Absorption units for modern systems include a generator and condenser in one shell, an evaporator and absorber in another, and the solution and evaporator pumps mounted below the two shells. Lithium bromide salt is used as the sorbent in many units. Here water acts as the refrigerant. The boiling point of the lithium bromide salt is so high that it behaves like a nonvolatile substance. In the above cycle there is no vaporization of absorbent in the generator and no carry-over of absorbent vapor to the condenser. The heat input to the cycle may come in the form of low-pressure (5 to 15 psig) steam or other fluids of comparable heat level. This factor opens many opportunities for utilization of process steam or waste heat—a unique advantage for industrial plants.

Absorption refrigeration has proved so practical and economical that these machines represent more than one-quarter of all refrigeration equipment sold with a capacity of 100 tons or more. Refinements in design have overcome nearly all limitations. The main exception has been the need for a separate cooling-tower bypass (Fig. 17-5a), and a condensing-water–temperature control valve has been developed recently.

Q Explain how this new control system (Fig. 17-5a) works.

A Conventional absorption systems require an external condenser-water temperature of around 85°F to keep the lithium bromide solution from salting up and keep the cycle of operation from drifting into the over-concentrated range. Thus the bypass is needed to achieve a constant entering water temperature. The two- or three-way control valve has had to be installed where it wouldn't freeze, and where it was accessible for repair and calibration.

In trying to stabilize its cycle, an absorption machine may undergo wild gyrations if the set point on the tower bypass drifts out of calibration, and the condenser-water temperature begins to fluctuate. Thus the cooling-tower bypass and steam valve "hunt" against each other, and the steam source is hard pressed to keep pace.

If the requirement for a fixed condenser-water temperature were eliminated, the absorbent solution could be cooled as the tower-water temperature dropped. The cooler solution would then be able to handle the load without expending excess steam in the generator. The part-load steam rate could be slashed because a cooler, more dilute solution would do the same job that a concentrated, hot solution ordinarily does.

This new control system does just that. Today, these new absorption machines in 100- to 1,120-ton sizes use uncontrolled condenser water down to 55°F to improve the part-load rate and reduce initial and operating equipment costs. A three-stage concentration control system replaces the bypass and valve. Now, by sensing solution temperatures and refrigerant water levels continuously, the system can quickly correct cycle imbalance by pumping refrigerant into the solution.

This advance prevents crystallization of solution under all tower-water–temperature conditions, while maintaining part-load performance. The steam consumption rate, which is about 18 lb per ton-hr at full load, drops to low as 15 lb per ton-hr at partial loads (see graph, Fig. 17-5b), taking advantage of the lowered condenser-water temperature as the load drops to as low as 15 lb per ton-hr at partial loads (see graph, Fig. 17-5b), taking advantage of the lowered condenser-water temperature as the load

Stable operation is maintained automatically from 100 per cent to slightly above 0 per cent by controlling load, steam rate, and leaving

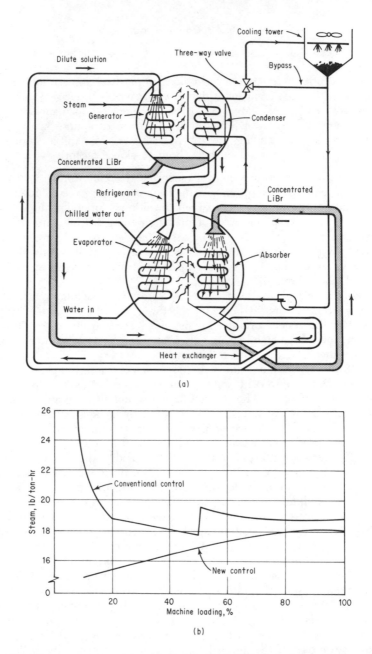

(a)

(b)

Fig. 17-5. Absorption systems (a) use heat instead of mechanical work to change refrigerant pressure. Sketch (b) shows comparison of steam required for various machine loads by conventional control and new control method developed in 1972.

236

chilled-water temperature. In comparison, conventional machines normally have to be shut off between 10 and 20 per cent load conditions.

Q Sketch a compression cycle, and explain how it works.

A See Fig. 17-6. The compression cycle remains our most familiar type of refrigeration for air conditioning. Essential parts of the compression system are the compressor, condenser, receiver, expansion valve, and evaporator or expansion coil. Compressors may be reciprocating or centrifugal and driven by electric-motor, turbine, or internal-combustion engines. Larger systems often use combination units for best efficiency over load range: centrifugal compressors, often turbine-driven, handle the base load while the system load swings are taken by reciprocating equipment. Combined operating efficiency cuts costs for plants with widely fluctuating cooling demands. Operating costs of the compression cycle make it a favorite of plants with high fuel costs. Refrigerants usually can be condensed by water or air at some handy temperature, say 80°F. This allows use of evaporative condensers, straight air-cooled condensers, or shell-and-tube water-cooled condensers. Incoming process water may be tapped for free cooling.

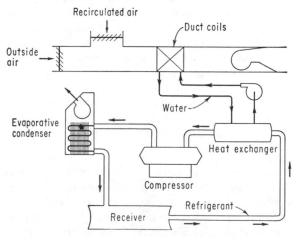

Fig. 17-6. Compression cycle handles most of our large cooling loads and is one of the most efficient systems.

Q Sketch a heat-pump cycle, and explain how it works.

A See Fig. 17-7. Heat pumps are a special brand of compression cycle used for the year-round air-conditioning system of the plant. During the cooling cycle the heat pump operates as in a straight compression-refrigeration cycle. Whenever heat is needed, the functions of the condensers and evaporator–heat exchanger are reversed. By rerouting flows

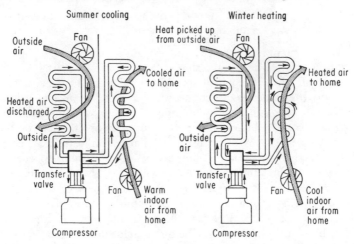

Fig. 17-7. Heat pump is a refrigeration system because refrigerant absorbs heat at one temperature and rejects it at a higher temperature.

the air-cooled condenser becomes the evaporator and picks up heat from the outside air. Extracting heat from 0°F outside air requires increased pressure drop in the refrigerant. New units often include a second compressor in the series to boost the pressure for heating. Again, centrifugal compressors may handle the base load with reciprocating units supplying the second compression stage. A packaged heat pump cuts the installation cost and permits use for spot-cooling operations or waste-heat reclamation. For most plants heat pumps are economical when the electrical rate is low and fuel costs high or when plant needs dictate simultaneous heating and cooling. You can often pump Btu's from one area to another for a reduction in air-conditioning load.

Q   How is the temperature controlled in an air-conditioning system?

A   Figure 17-8 shows a typical air-conditioning system for a large plant. The temperature is kept at a set, comfortable level by the addition or removal of heat. To do this, two heat exchangers are used—one for heating in winter, the other for cooling in summer. The heating medium may be hot water or flue gas. Steam or hot water flows through coils placed in the air stream to be heated. The temperature is usually regulated by controlling the steam or water temperature in the coil with a thermostat. Another way to regulate the temperature is to pass unconditioned air past the damper, thereby mixing heated and unheated air. The cooling medium is chilled water, brine, or a direct-expanding refrigerant. Cooling is also done by passing unconditioned air through cold-water sprays. Outdoor air is heated by steam coils, then humidified or dehumidified as needed. After the air is washed it is distributed by a fan through ducts.

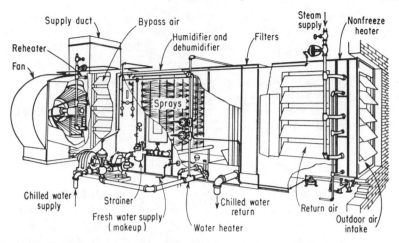

Fig. 17-8. Central plant has all conditioning elements except refrigeration.

Q How do human beings affect the air they live in?

A Human beings (1) use up oxygen in metabolism, (2) exhale carbon dioxide, a waste product of metabolism, (3) increase the dry-bulb temperature of the air because energy liberated in metabolism is partly lost as heat, (4) increase the humidity of the air because of the evaporation of moisture from their body surfaces and through respiration, (5) give off odors to the air, and (6) decrease the number of small ions in a unit volume of air.

Q How are odors removed from conditioned spaces?

A See Fig. 17-9. Activated carbon, the same air-filtering agent used in gas masks, removes odors from normal manufacturing and commercial operations. When it is used to purify recirculated air, activated-carbon air recovery supplants the need for new outdoor air. It is less costly to recirculate or reuse air that is already at the proper temperature and humidity than to heat or cool new outdoor air. Activated carbon comes in canisters or in panels. Both types are placed in ducts, casings, or plenums. Air-recovery units should always be protected by dry-type dust filters to prevent carbon clogging. Carbon is placed either in return air ducts from conditioned

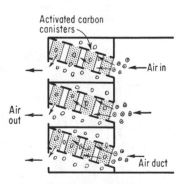

Fig. 17-9. Canisters of activated carbon are placed in the air stream.

zones or in discharge vents from kitchen ranges, etc., to keep them from

240   Standard Refrigeration and Air Conditioning Q & A

discharging odors to the neighborhood. Removing odors, dust, fumes, pollens, and other contaminants all come under air purification. In general this is done by placing filters, activated carbon, washers, etc., in the air stream.

Q   What are some causes of odors in conditioned spaces?

A   Odors in conditioned areas may be caused by refrigerant leaks, brine leaks, no trap in the air-washer drain to the sewer, a dirty cooling-coil surface, or odor-producing units located near the air intake.

Q   What is spray humidifying?

A   Spray humidifying is the process of adding moisture to the air by passing it through water sprays (washers). The sensible heat of air changes into latent heat, which supplies the heat needed for vaporizing some of the water. Though the dry-bulb temperature of the air goes down,

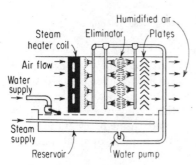

Fig. 17-10. Air washer for air-conditioning system has these components.

the total heat of the air doesn't change (only proportions of sensible and latent heat change). Figure 17-10 shows a typical air washer. The air flows through filters that separate dust from the air stream and then to the water sprays that remove the remaining impurities by washing.

Q   How are dust particles removed from conditioned air?

A   See Fig. 17-11a to c. The devices used for air purification include (1) centrifugal collectors, (2) air washers, (3) dry-type filters, (4) viscous filters for manual operation, (5) viscous filters for automatic operation, (6) electrostatic precipitators, and (7) absorption equipment. Cen-

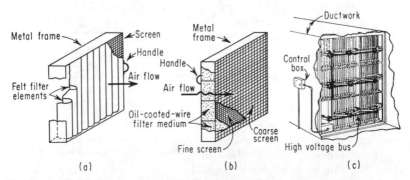

Fig. 17-11. Air filters in various designs are placed in the air stream.

trifugal collectors remove heavy airborne contaminations and cope with the waste from woodworking machines, the exhaust from sanding and grinding, the fly ash, etc. Air washers are banks of spray nozzles that arrest entrained particles. Surface scrubbers have large surfaces that are constantly washed down by spray nozzles. Impurities are removed with spray. Dry-type filters are either throwaway or semipermanent. Throwaway types are discarded when saturated with airborne impurities. Semipermanent filters are cleaned and used again a number of times. Viscous filters are usually made of wool or fiber, screen wire, metal punchings, etc., and are treated with a sticky oil that captures and holds particles that flow through the air. Electrostatic precipitators arrest airborne particles under the influence of a high-voltage electric field. The particles become charged or ionized and will precipitate when they contact an electrode of the opposite polarity. Absorption equipment is used to absorb odors, organic gases, and vapors.

Q How are air mixtures handled in air-conditioning systems?
A In many systems, the air returning from the conditioned space is mixed with incoming fresh air. In most systems, a certain quantity of air is discharged from the system and replaced with fresh or new air. The balance of the air is recirculated through the conditioner. It is recooled or reheated and humidified or dehumidified. In systems where the contaminated air is very slight, recirculated air isn't cleaned but is mixed with fresh air that is cleaned before it enters the system. Where contamination is heavy, both fresh and recirculated air are treated. In some systems, even the air wasted must be cleaned before it may be discharged to atmosphere. Fresh air is air taken from out-of-doors to replace the air discharged from the system. The degree of freshness may vary widely; it may even be less pure than the air discharged from the conditioned system. Therefore cleaning is important.

Q What are packaged air-conditioning units?
A See Fig. 17-12. Packaged air-conditioning units come with all the needed equipment in a single cabinet. They are usually made in smaller sizes and placed in various zones around a plant or office. Large packaged units have ductwork for fully automatic room conditioning that is too large for one outlet.

Q Name six important steps to consider when planning air conditioning for an old building.
A (1) Locate the equipment room as near the load as is possible; (2) allow space not only for machines but for servicing; (3) choose equipment that is easiest to bring through the building entrance; (4) make sure the electrical service has the capacity to carry the load; (5) provide sufficient water and water conservation devices; and (6) install drainage facilities if they are not already in the machine room.

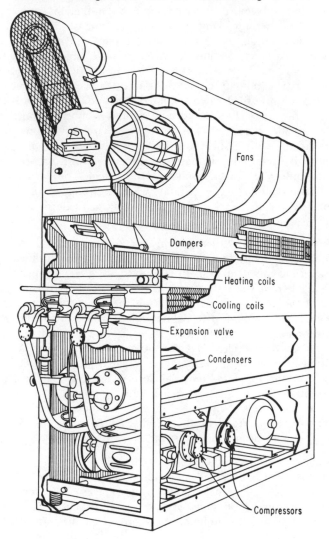

Fans

Dampers

Heating coils

Cooling coils

Expansion valve

Condensers

Compressors

Fig. 17-12. Packaged air-conditioning unit is complete in one package.

Q   What do codes and sanitary regulations usually require in drainage?
A   Codes and sanitary regulations usually require that equipment drains discharge into a water-supplied open sink, properly vented and trapped. Figure 17-13 shows a typical installation with a ceiling-type air-handling unit. It was used in this installation so the condensate could drain by gravity flow from the drip pan to the sink. If a machinery room doesn't have a floor drain, elevate the condenser so it can be drained when serv-

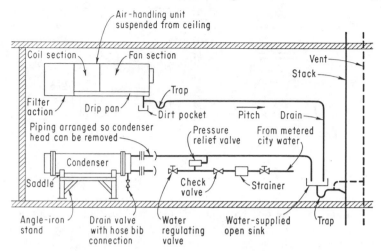

Fig. 17-13. Plan drains for servicing suspended ceiling unit and condenser where floor drains can't be used.

iced. Condensers do have to be opened for cleaning and even for replacing tubes at times.

Q What do some city codes say about water conservation?

A In one city, for example, any installation consuming more than 2,650,000 gal of water per year, which is about what an average 35-ton system uses in one summer, is required to have a water-conserving device, such as an evaporative condenser or cooling tower. Air-conditioning systems need about 1½ gpm of cooling water per ton of refrigeration. The New York City code requires water conservation in all systems of 6 tons and over.

Q Are refrigerants allowed to be expanded directly into the cooling coils installed in air-conditioned spaces?

A Local codes permit certain refrigerants (Group 1, American Standards Association) to be used directly in the cooling coils. In such direct-expansion coils the liquid refrigerant vaporizes and in so doing extracts heat from the passing air stream.

Q What is the ice system of air conditioning?

A See Fig. 17-14. In this system cooling is done by melting ice. This is a practical air-conditioning method for churches, theaters, and public halls that have short operating hours and relatively high peak loads. Since the investment in mechanical refrigeration equipment is expensive for short periods, ice fills the need. A small quantity of ice in a water-cooling tank can release refrigeration at a rapid rate.

Q How can natural ground water be used in air conditioning?

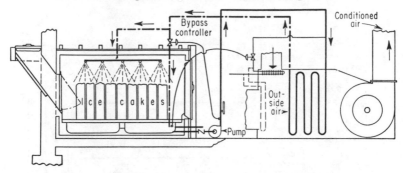

Fig. 17-14. Ice system of air conditioning uses ice as the cooling medium.

A Some ground water remains at a low constant temperature of around 55°F or cooler. This water can be pumped through cooling coils to pre-cool the air before it is cooled by the chilled refrigerated coils. This slightly heated water is then returned to a disposal well some distance away to conserve ground water.

Q How is air blended?
A See Fig. 17-15. To blend air, an air washer in front of the supply fan is bypassed in hot weather. The supply fan discharges through a split duct; the top is a heating coil and the bottom is a cooling coil. The air leaving these coils goes into separate chambers, with branch ducts running from each chamber. Hot and cold ducts run side by side. At each take-off there is a double-bladed mixing damper control, a room thermostat, or other control device. This blends warmer and cooler air as needed to give the desired conditions in each space. In the winter this system does not use a refrigerant in the cooling coil; in the summer there is no heat in the heating coil. Because air leaving the warm chamber may be high in relative humidity, it may not compensate for the nearly saturated cool air. Closer control over the relative humidity may be obtained by placing a reheater in a cold-air chamber for summer service.

Q What is a central air-conditioning system?

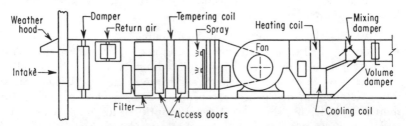

Fig. 17-15. Air can be supplied at two different temperatures with this system of blending.

A A central air-conditioning system has all of the major equipment—filters, air washers, fans, and refrigeration machinery—in one centrally located space, removed from the area to be conditioned. Ducts distribute the conditioned air to the desired spaces.

Q What are some advantages of a central air-conditioning system?

A (1) The space occupied by the equipment need not be as valuable as the conditioned spaces. (2) For a large conditioning load the equipment may cost less. (3) The maintenance and inspection of a central system does not disturb the people in the conditioned areas. (4) The exhaust air can be returned and partly reused with obvious savings in heating and refrigeration.

Q What is zoning?

A Zoning is used to cope with different conditions in various parts of large buildings or plants. For example, more air conditioning may be needed on the sunny side of a building than on the shady side. Thus, areas are broken up into zones, and each is fed by separate ducts.

Q What is a high-velocity air-conditioning system?

A See Fig. 17-16. To save space and installation cost, air can be forced through relatively small conduits at high speeds. Here, high-velocity conditioned air acts as a jet to induce room air with the incoming conditioned air. The mixture sweeps over a heating or cooling coil for final conditioning. The water supply and return lines leading to these coils run parallel to the air conduits, which are furred-in along the outside walls.

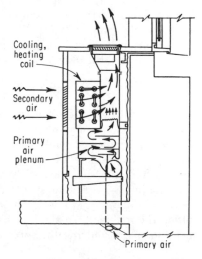

Fig. 17-16. High-velocity air-conditioning unit uses smaller conduits.

Q What is reverse operation?

A Room conditioners having refrigeration equipment inside a room may be used for heating during the winter months. Under these conditions the cooling coil becomes the condenser and the condenser acts as the evaporator. In this way the unit abstracts heat from the outside air and delivers air at a higher level to the room. This is similar in operation to a heat pump.

Q How do you control refrigeration compressor capacity?

A   Compressor capacity is controlled by the unloader or bypass valves. Most compressors have two to three steps; or if there is more than one compressor, they may be started or stopped in any order to give the required capacity. Multiple-compressor jobs may use bypass valves on each compressor.

Q   How do you prevent vibration from traveling through the ducts?
A   Vibration in ductwork can be minimized by setting all air-conditioning equipment on vibration isolators. Separate the equipment foundations from the main building. Install sound-absorbing linings inside the ducts and air diffusers, and keep the air diffusers tight against the ducts to prevent air leakage; leakage also adds noise to the system. Air diffusers placed in the ceiling are quieter than those located in the walls, even if the wall outlet is below the ceiling. Separating the ducts of both supply and return lines prevents noise from traveling from one area to another. When passing ducts through the walls, you should insulate them at the cross-through points to avoid wall noise pickups. Floating-construction ducts mounting also eliminates duct noise. Keep the duct elbows from branching too close to the fan outlets.

Q   How are drafts avoided in conditioned spaces?
A   To avoid drafts in conditioned spaces, use ceiling diffusers with bottom grille dampers. With this system, the air velocity can approach 3,000 fpm and will be diffused with lowered noise in spite of the machines, columns, and other obstacles.

Q   What is a sprayed roof, and how does it help an air-conditioning system?
A   Large flat roofs are often sprayed by jets or nozzles to intercept the sun's rays before they reach the roof. The nozzle sprays are thermostatically controlled, while the rotary jets use no thermostat or electrical wiring. The advantage of a sprayed roof is that top floors that take a beating from the sun are cooler, and as much as 20 per cent less air conditioning is needed in spaces directly under them. A sprayed roof cools in two ways: (1) water droplets intercept the sun's heat before hitting the roof and (2) the roof film evaporates, pulling heat out of the roof and spaces underneath. Sprayed roofs are much cooler than flooded roofs.

BOILERS USED IN AIR-CONDITIONING SYSTEMS

NOTE: The subject of boilers is fully covered in *Standard Boiler Operators' Questions & Answers*, 1969, which includes the latest ASME Codes, safe operating practices, etc.

CALCULATIONS FOR AIR-CONDITIONING TROUBLESHOOTING

Q Give a quick check list of temperatures, psig, gpm per ton, and pressures handy when troubleshooting air-conditioning refrigeration systems.

A Fig. 17-17 gives typical ranges useful for many such jobs.

Q In troubleshooting, what is the advantage of knowing the capacity of a system that cools chilled water for air conditioning?

A Most such systems have a water chiller in which refrigerant expands in a coil. Water passes through the chiller shell. But the same method can be used for a flooded-type chiller. Figure the capacity thus:

$$C = 0.024 \times gpm \times t_a$$

where C = tons of refrigeration

gpm = gallons of chilled water circulated per minute

t_a = temperature (°F) difference between water entering and leaving the system's chiller

If there is no flowmeter to measure chilled-water flow, just multiply the difference between the suction and discharge pressure-gage readings of the circulating pump by 2.31. This gives the pump head in feet. Refer to the characteristic curve for your pump and read the number of gpm corresponding to that head. Fig. 17-18 shows a typical curve (don't use this for figuring).

To estimate gpm from the curve, you must know both the head and the horsepower of your pump. The easiest way to find pump hp is to use a clamp-on wattmeter to measure the input wattage. Divide the reading

Item	Range
Condensing-water inlet temp..............	60–95°F
Condensing-water outlet temp.............	85–115°F
Water-temp rise in condenser.............	15–25°F
Freon-12 head pressure..................	100–150 psig
Freon-12 suction pressure................	25–50 psig
Temp air enters cooling coil..............	75–90°F
Temp drop through cooling coil............	12–25°F
Temp air leaves cooling coil..............	50–70°F
Air circulated per ton refrigeration.........	300–600 cfm
Room temp for comfort cooling............	70–86°F
Condensing water......................	3–5 gpm per ton
Temp chilled-water enters coolers..........	50–65°F
Temp chilled-water leaves coolers..........	40–50°F

Fig. 17-17. Quick check list for air-conditioning refrigeration systems.

by 746, multiply the result by 0.80 and the pump efficiency, then read from the characteristic curve. The value 0.80 is the assumed motor efficiency, 80 per cent, and is expressed as a decimal.

EXAMPLE: A compressor chills 100 gpm of water through a temperature range of 20°F. What is the compressor capacity in this load?

SOLUTION: Use the first equation above.

$$C = 0.024 \times 100 \times 20 = 48 \text{ tons}$$

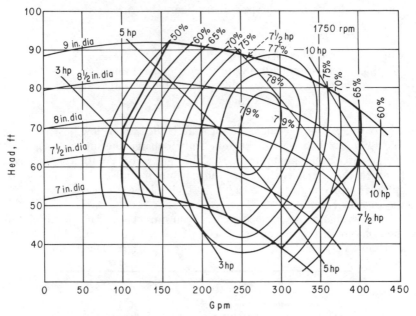

Fig. 17-18. Typical characteristic curve for centrifugal pump in air-conditioning system. Don't use for calculations.

Q Show how to calculate the capacity of direct-expansion systems.

A The over-all capacity of direct-expansion systems may be figured in three ways: (1) measure the total heat removed from air passing through the coil, (2) measure the heat rejected to the condensing water, and (3) separately measure sensible and latent heat removed from air.

Sensible-heat removal results in a change in air temperature, while latent heat comes from moisture that condenses out of the air. Method 2 is equally good for direct and indirect systems.

Q How is total heat figured?

A To figure total heat removed from air passing over a coil, use this equation:

$$H = 4.5 \times \text{cfm} \times h_a$$

where H = total heat removed from air in Btu per hr
   cfm = cubic feet of air passing over the coil per minute
   h_a = enthalpy difference between entering and leaving air in Btu per lb

Obtain air enthalpy from a psychrometric chart or from the table in Fig. 17-19. Read enthalpy at the air's wet-bulb temperature, not the dry-bulb temperature. Once you figure total heat removed by the refrigerant, convert this to tons of refrigeration by dividing by 12,000 Btu/(hr) (ton). If the wet-bulb temperatures used are above or below those in the table, see a standard reference book, such as the ASHRAE Guide.

EXAMPLE: Air enters a coil at a wet-bulb temperature of 80°F and leaves at 70°F. What is the compressor tonnage when 1,000 cfm flow through the coil?

SOLUTION: The difference in total heat, using values from Fig. 17-19, is 42.64 − 33.51 = 9.13 Btu per lb. Using this in the equation to find the total heat removed, $H = 4.5 \times 1,000 \times 9.13 = 41,085$ Btu per hr. This is 41,085/12,000 = 3.43 tons of refrigeration.

Wet-bulb temp, F	Enthalpy, Btu per lb
68	31.92
69	32.71
70	33.51
71	34.33
72	35.17
73	36.03
74	36.91
75	37.81
76	38.73
77	39.67
78	40.64
79	41.63
80	42.64
81	43.67
82	44.72
83	45.80
84	46.91
85	48.04

Fig. 17-19. Table showing enthalpy of moist air.

Q   How is air velocity measured in an air-conditioning system?
A   Use a meter if possible, as it gives velocity readings directly. Convert to cfm by multiplying the average velocity as obtained from a dozen or so readings across the duct by the duct area in sq ft. You can also use a pitot tube to obtain air velocity; this is handy when a duct opening is not accessible, but be sure to follow the instructions that come with these instruments when taking measurements.

Q   How is heat rejected to the condensing water figured?
A   Figure from

$$h = 500 \times \text{gpm} \times t_a$$

where h = heat to condenser water, Btu per hr
    gpm = gallons of water passing through condenser per min
    t_a = temperature (°F) rise of water through condenser

Deduct 10 per cent of the figured heat to get the actual heat rejected because this much of the heat in the condenser water is what is known as "heat of compression." This is heat from work that the compressor must do to raise the pressure of a vapor. Divide the result by 12,000 to convert to tons of refrigeration.

Also, if the compressor is of the hermetic type, with the motor winding cooled by suction gas, this heat is also rejected to the condenser water. Thus deduct the Btu equivalent of motor heat, or motor hp × 2,544, from the heat load on the condenser. Treat fan motors in the cool-air stream the same way.

Remember, this figuring holds for all shell-and-tube condensers. Use a flowmeter or measuring barrel to find water flow in gpm. On large systems where it is difficult to measure flow accurately with a barrel, install pressure gages on the inlet and outlet of the condenser water pump. Then multiply the difference in readings by 2.31 to find the total head in feet. Read the gpm from your pump's characteristic curve, after figuring motor hp as explained above.

Q   Can the capacity of evaporative condensers be figured by the method above?
A   Yes, but now we are dealing with an air-temperature rise instead of a drop. Also, you must subtract the heat of compression (10 per cent) and the electrical heat, if any.

Total sensible heat is found from

$$h_a = 1.1 \times \text{cfm} \times t_a$$

where h_a = total sensible heat removed from air in Btu per hr
    cfm = cu ft of air cooled per min by refrigerant
        t_a = temperature difference between air entering and leaving coil
            in °F

Figure the total latent heat from

$$h_1 = 8,750 \times \text{gph}$$

where h_1 = total latent heat removed from air, Btu per hr
gph = gallons of moisture condensed from air per hr

Add total sensible and latent heat to find the amount of heat the coil removes from air.

EXAMPLE: What is the capacity of a refrigeration unit using 30 gpm of condensing water when the temperature rise through the condenser is 20°F?

SOLUTION: Using method 2, $h = 500 \times 30 \times 20 = 300,000$ Btu per hr. Subtracting 10 per cent for heat of compression, the quantity of heat rejected to the cooling water is 270,000 Btu per hr. The tonnage is $270,000/12,000 = 22.5$ tons.

Q Where should room thermostats be located for best results in a heating-cooling system?
A For best over-all results, the control should be in the space which has the most occupants. A single large space may need several controls. A building doesn't have to be very large before good control dictates at least one zone for each exposure, especially when cooling is included. In winter, for example, with prevailing northwest winds, it may be possible to satisfactorily combine the north with the west zone, and the east with the south. If there are partitions in the space, a good spot is on a partition a few feet from the outside wall.

Q Is it economical to turn back thermostats on heating systems?
A According to the Heating, Ventilating, Air Conditioning Guide, Vol. 38, 1960: "Though opinions vary regarding the amount of fuel that can be saved by automatic night set-back, tests made at the University of Illinois indicate that, on thermostatically controlled systems, a possible fuel saving from 7 to 10 per cent may be obtained by reducing the building temperature 6 to 10°F from about 10 PM to 5:30 AM."

STARTING AIR-CONDITIONING CENTRIFUGAL COMPRESSOR SYSTEMS

Q Name twelve basic steps for starting centrifugal systems.
A There are various types of systems and controls, but here are basic steps for starting: (1) Check oil levels in the compressor, drive, gear, and coupling to make sure they are all right. (2) Start condenser-water flow. Be sure to avoid water hammer in the system. (3) Start brine circulating through the brine cooler. Again, avoid water hammer. (4) Check the air pressure of air-operated controls, if any. (5) Run the purge unit to rid the machine of air. Always do this before starting. (6) Close the

suction damper only as far as necessary on synchronous-motor drives. (7) Warm the turbine on turbine-driven machines. Be sure to drain the system thoroughly before starting the turbine. (8) Close the holding circuit for safety controls, if necessary for starting. (9) Bring the machine to full speed and be sure the seal-oil gages have the right oil pressure. (10) Open the air supply to the controller on automatically controlled machines. (11) Open the supply valve in the water line to the oil cooler on the unit for drive and reduction gears. (12) Run at high speed if machine surges. This accelerates purging.

Q   Explain how to start a purge-recovery unit.
A   Run a purge-recovery unit either continually while the machine operates or when a purge shows it's needed. After starting synchronous-motor-driven units, open the suction damper as required as soon as the compressor reaches full speed.

Q   Give details on starting a steam turbine drive.
A   Turbine starting depends on the turbine make, each design is somewhat different. Adjust the governor control valve to its minimum setting, then make sure all drains blow steam before starting. Bring the turbine to full speed. If seal-oil pressure doesn't develop within a few seconds after starting, stop the machine and restart. On automatically controlled machines, opening the air supply to the controller brings the machine under the supervision of the controls.

Keep the water flow to the compressor oil cooler at a low rate until the highest bearing temperature is about 130°F. Adjust to give a bearing temperature of 140 to 180°F, or whatever value the builder recommends. Check the temperatures of all bearings on the compressor, gear, and drive until they level off at the right point.

Q   Explain surging operation of centrifugal compressors.
A   Surging operation is normal for centrifugal compressors. There's nothing to worry about. But at low loads (10 to 20 per cent of full load), surging causes compressor overheating and the bearings get hotter. So don't operate the compressor continuously under these conditions. If you want to operate continuously at light loads, a hot-gas bypass may be needed.

Shortly after starting there may be a period of surging until all air is removed from the condenser. Meanwhile, run the machine at high speed. But condenser pressure shouldn't exceed 15 psig for Freon-11 during the high-speed period. Also, the input current to the motor of a motor-driven unit shouldn't be more than 100 per cent of full-load motor rating. And don't overcool the evaporator, because the nonfreeze control will stop the machine. Adjust the speed or the damper to give the right brine temperature after the machine has steadied and all the air is purged from the system.

Q How do you use hand operating controls on startup?

A Three methods for hand-controlling the capacity of a centrifugal compressor are available: (1) vary the compressor speed, (2) throttle the compressor suction, or (3) increase the compressor discharge pressure. The first method is most efficient, the third least.

You can change the speed at the driver, whether it is motor or turbine drive, except with constant-speed motors using magnetic slip or hydraulic coupling. The throttling damper in the cooler suction connection is used for suction-damper control. Throttling the suction increases the pressure range through which the compressor must handle refrigerant vapor. Since the machine has a flat characteristic curve, the pressure range against which it will deliver is limited by the operating speed, so suction throttling is a good method of control. It uses a little more power at partial loads, but it gives stable (nonsurging) operation at lower loads than variable-speed control. Some machines are fitted with both variable-speed and damper control.

Q Explain condenser-water control of centrifugal units.

A Condenser-water control isn't often used, but has its good points for machines taking condensing water from city mains or other high-cost sources. When you throttle condenser-water flow you get the same action you would get with a suction damper. On some jobs the lowest speed you can get with variable-speed control may not be low enough to meet your operating needs. On these jobs you may throttle condenser water and bring the compressor speed into the range of control.

Q Can the hand controls listed in previous questions be automated?

A Yes, Figs. 17-20 and 21 show setups for automatic variable-speed control of turbine-driven units. The systems automatically adjust the compressor speed to hold the delivered brine temperature constant. A rise or fall in brine temperature works through the controller to supply air or bleed it from a pneumatic valve in the turbine-governor system. This valve opens or closes the governor steam valve to increase or decrease the turbine speed.

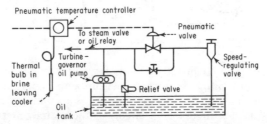

Fig. 17-20. Automatic variable-speed control for turbine-driven compressor.

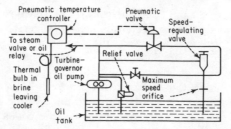

Fig. 17-21. Another setup for automatic variable-speed turbine-driven compressor.

Q How does an automatic damper control on a centrifugal system work?
A Fig. 17-22 shows an automatic damper control. The temperature controller regulates the degree of damper throttling by supplying or bleeding air from the pneumatic damper motor. The condenser-water control, Fig. 17-23, uses a diaphragm valve in the water discharge. The valve is automatically positioned in response to changes in delivered brine temperature. A reverse-acting temperature controller supplies or bleeds air from the valve diaphragm. If water is scarce or expensive, a similar control which has a bulb in the water leaving the condenser can be used.

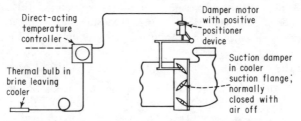

Fig. 17-22. Automatic damper control regulates throttling in suction.

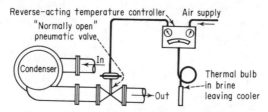

Fig. 17-23. Automatic condenser-water control uses a diaphragm valve.

Fig. 17-24 show control methods for motor-driven units with magnetic or hydraulic clutches. In small machines an excitation-control rheostat used with a magnetic clutch is in the d-c supply to slip couplings. Bigger units have it in the field circuit of the motor-generator set.

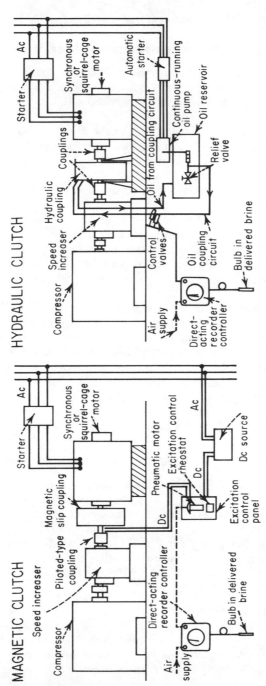

Fig. 17-24. Automatic variable-speed controls of motor-driven centrifugal compressors use above type of couplings. Temperature-sensitive bulb in brine cooler discharge actuates control to change speed.

255

SHUTDOWN PERIODS OF CENTRIFUGAL REFRIGERATION UNITS

Q   There are two types of shutdown periods: (1) standby shutdown, when the machine must be ready for instant use, and (2) extended shutdown, when the machine is out of service for the season. How is the first one handled?

A   During a standby shutdown: (1) Run the purge-recovery compressor as needed to keep the pressure in the machine slightly below that of the atmosphere. (2) Keep the machine free of air leaks. (3) If machine pressure builds up as a result of a warm machine room rather than air leakage into the unit, circulate a small quantity of water at less than 70°F through the cooler or condenser. Do this either intermittently or continuously. Or operate the purge-recovery unit to pull the pressure down. (4) Run machine a few minutes once a week to circulate oil and reduce refrigerant temperature. Be sure refrigerant temperature in the cooler is not lowered below the freezing point of the water or brine in the tubes. (5) Keep the machine room above freezing temperature. (6) Keep the right oil level.

For long shutdowns, if the system is known to be free of leaks and the purge-recovery unit holds machine pressure down to the proper value, follow the instructions for standby shutdown. Also: (1) If freezing temperatures are likely at the machine location, drain all the water from the compressor, gear, and turbine-oil coolers. Then drain the water from the condenser, cooler, seal jacket, pumps, piping, and other units. (2) During a long shutdown, oil may be excessively diluted by refrigerant. If the oil level in the pump chamber rises into the rear bearing chamber, remove the entire oil charge and tag the motor switch or turbine.

SHUTTING DOWN THE AIR-CONDITIONING SYSTEM FOR THE SEASON

Q   Explain how to put your air-conditioning system "to bed" for the season.

A   Never merely turn the switch off and forget about the equipment until it is needed next spring. Here are ten things to do: (1) check the system for leaks, (2) pump down the system, (3) drain cooling water, (4) remove scale, (5) clean tubes, (6) slack off belts, (7) clean the crankcase, (8) renew crankcase oil, (9) clean the suction strainer, and (10) disconnect the starting circuit.

Q   How is a system pumped down?

A   As soon as the system has been tested for leaks (use at least 50 psi) and made tight, it can be pumped down. This means locking up all refrigerant in the receiver, or in the condenser if no receiver is installed. This avoids losing refrigerant during the idle season, and also keeps excessive pressure from developing in the low side if a heating coil is near the evaporator. High pressure often blows out the compressor seal.

Close the main liquid valve, then set the room thermostat to a temperature low enough to cause the solenoid valve to open. Be sure the fan is running during this operation. On many systems the solenoid valve is energized from the load side of the fan starter to prevent mistakes. Start the compressor and run it until all liquid is boiled out of the evaporator as well as the liquid line.

If the low-pressure control stops the machine before the suction pressure drops to 2 psig, hold the control by hand until this pressure is reached. The pressure will rise after the compressor stops, due to refrigerant boiling out of the crankcase oil.

Wait several minutes, then trip the low-pressure switch by hand and pump down to 2 psi again. Repeat as required until the pressure stays at 2 psi. This removes all liquid refrigerant from the oil in the compressor and from traps in the system. If 2 psi cannot be reached, look for leak-back from discharge, which will usually come from compressor head valves. Then close valves A and B, Fig. 17-25, to isolate the liquid in the receiver until it is needed next season.

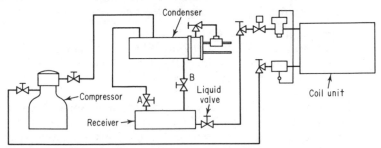

Fig. 17-25. When shutting down for season, close valves A and B to lock refrigerant in receiver.

Q What attention must be given to a shell-and-tube condenser?
A Shut off the water supply, drain the water side, remove heads, and inspect the tubes. If scale is present, clean the tubes with a wire brush sized properly for a tight fit.

CAUTION: Don't scratch the tubes, as this often starts erosion which results in tube failure.

If a wire brush does not remove scale, use a good industrial scale solvent. Most of these are inhibited acids. Observe the manufacturer's directions closely and be sure to follow the acid cleaning with a neutralizing wash with an alkaline solution. Soda ash is a good source of alkali. Flush the condenser once more with clean water.

Heads may be bolted on again, but leave the drain plugs out. Disassemble and clean the water-regulating valve completely. Be careful to shut off the gas connection to the valve before starting work on it. Remove the handwheel of the water shutoff valve, or tag it to avoid its being opened by mistake.

Q   What must be done to secure evaporative condensers?

A   Drain out water and clean the sump pan thoroughly. If the galvanizing is damaged, paint the pan with a good metal primer. Disassemble and clean all spray nozzles. The tube bundle in most evaporative condensers is too compact to permit effective manual cleaning.

If the tubes are badly scaled, a chemical cleaner may be circulated through the spray system. Neutralize and rinse as outlined for shell-and-tube condensers. Carry out this operation before painting the sump pan. Be sure to drain the circulating pump casing to prevent freezing. Take the steps listed above to prevent accidental opening of the water supply valve.

Q   How is the water tower secured for the season?

A   The same steps that apply to evaporative condensers apply here, except for tube cleaning. Redwood-constructed towers need no painting.

Q   How are compressors secured for the off-season?

A   Slack off belts to reduce stretching. Drain out crankcase oil and clean the crankcase, if it is accessible. Refill it with clean oil. Close both the suction and discharge service valves. Clean the suction strainer, if one has been installed.

Some engineers take the fuses out of the compressor motor circuit to prevent accidental starting with the compressor valves closed. But you should go a step farther and disconnect a wire in the starting circuit. One lead to the low-pressure cutout is the most common choice. Pull this lead free of the contacts and tape the terminal well. A note placed in the fuse box will be a helpful reminder next spring.

Q   How do you set up an air-conditioning system for heating?

A   Some air-conditioning systems use steam for heating. After securing the refrigeration system as outlined above, there are only a few steps in preparing for heating.

Check the filters to see that they are clean. Set the manual outside air damper to the *winter* position. Most systems are designed to use less outside air in winter than in summer. Don't block the automatic damper to accomplish this, as it will damage the motor.

Check the steam-supply and condensate-return lines for tightness.

Set the controls for heating and see that the transformer, thermostat, and steam valve operate properly.

Follow all these steps religiously and your air-conditioning system will be raring to go when called on next spring and will start off efficiently when it is needed.

COMPUTERIZED CONTROL FOR A-C SYSTEMS

Q Explain how the computer fits into today's a-c control system.

A Fig. 17-26 shows schematically a computerized control center. The computer and the usual control center functions are combined in one package, but are shown separately for clarity. The computer memory (details given in *Standard Instrumentation Questions and Answers*) stores local weather records, the calendar, and expected schedules of building operation. Based on data in its memory, the computer takes current weather and interior conditions into account in order to determine the optimum supply temperatures and flows for lowest-cost operation of the system.

Equipment operation follows a predetermined procedure, also stored in the memory. Schedule revisions, desired space condition changes,

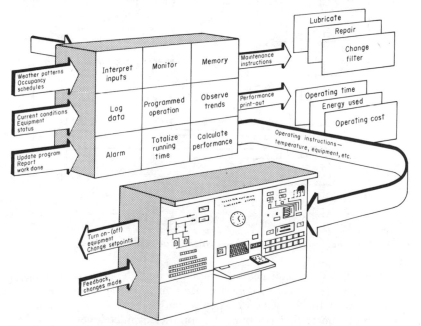

Fig. 17-26. Computerized control center for air-conditioning system.

equipment repair reports, etc., are continually fed into the computer to update its programming. Computer outputs include maintenance instructions based on equipment running time or deteriorating efficiency.

Q   How does the control center start the refrigeration compressor?

A   The signal to start the compressor comes from the control center, Fig. 17-27 (1), and activates the compressor control panel (2). If chilled water is flowing, the cycle continues. Condenser-water and auxiliary oil pumps start under the direction of the panel. When all conditions are satisfied, including closed inlet vanes (permitting an unloaded start), the motor starter is energized.

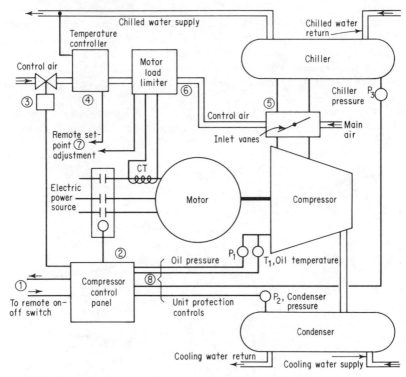

Fig. 17-27. Control hookup for refrigeration system of air-conditioning unit.

After the start sequence is completed and the unit is running, a solenoid valve (3) opens to deliver control air to the load controller (4), which controls the chilled-water-supply temperature. Control air moves a pneumatic motor that positions the compressor inlet vanes (5). When more cooling is needed, the controller raises the air pressure to open the vanes. A motor load limiter (6) puts a ceiling on control pressure so that the compressor cannot exceed the motor peak load.

The control center directs performance with remote set-point adjustments (7) for the controller and load limiter. Malfunction indicators (8) shut down the machine and trip annunciators in case of trouble.

Q How does the computer start the air-handling system?

A A switch at the control center starts the supply and return fans (Fig. 17-28) (1), opens the minimum outdoor and exhaust air dampers (2), and activates the control system. The outdoor air controller (3) is an on-off device that closes the maximum air dampers (4) whenever outdoor air cannot provide any cooling.

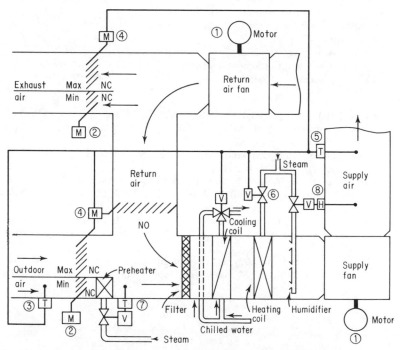

Fig. 17-28. Automatic air-handling systems supply ventilation needs.

A temperature controller (5) maintains the supply temperature. In summer it controls chilled water. When the outdoor temperature falls below the outdoor controller set point, the maximum and return dampers switch to supply the controller (5). Outdoor and return air mix to maintain the temperature until maximum dampers are again closed.

A further temperature drop opens the steam valve (6). The preheat controller (7) sends steam to the preheat coil when there is any danger of freezing. The humidity controller (8) adds moisture through steam jets in winter.

The control center receives various temperatures, flows, and damper positions. Set points and damper positions can be revised from the control panel in the control center of the building.

SUGGESTED READING

Reprints sold by *Power* magazine, costing $1.20 or less
    Power Handbook, 64 pages
    Air Conditioning, 20 pages
    Conditioned Air for Industry, 16 pages
    Refrigeration, 20 pages
    Handbook on Fans, 16 pages
    Balancing Rotating Equipment, 24 pages
    Vibration Isolation, 16 pages

Books
    Elonka, Steve, and Joseph F. Robinson: *Standard Plant Operator's Questions and Answers*, vols. I and II, McGraw-Hill Book Company, New York, 1959.
    Elonka, Steve: *Plant Operators' Manual*, rev. ed., McGraw-Hill Book Company, New York, 1965.
    Elonka Steve, and Alonzo R. Parsons: *Standard Instrumentation Questions and Answers*, vols. I and II, McGraw-Hill Book Company, New York, 1962.
    Elonka Steve, and Julian L. Bernstein: *Standard Electronics Questions and Answers*, vols. I and II, McGraw-Hill Book Company, New York, 1964.
    Elonka, Steve, and Anthony L. Kohan: *Standard Boiler Operators' Questions and Answers*, McGraw-Hill Book Company, New York, 1969.

18

COOLING WATER AND BRINE

Q   Is there an ideal treatment for all cooling water?

A   Economics plays an important role in the selection both of cooling-water systems and of methods of water treatment. It is possible to prescribe an ideal water for each cooling job, but the expense of the required equipment and chemicals is out of the question for some installations. Therefore, treatment must be tailored to fit the size and type of cooling-water system currently in use.

Q   What are the main types of cooling systems?

A   The two main classifications are "once through" and "circulating." In the simple once-through system water passes through the heat-exchange equipment only once. It is increased in temperature about 20°F before going to waste. Circulating systems are more complex (Fig. 18-1). Cool-

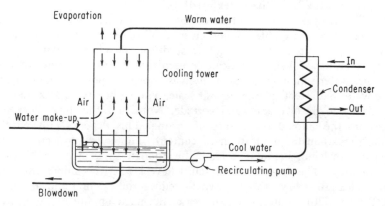

Fig. 18-1. Open recirculating hookup makes use of cooling tower or spray pond to cool water by evaporation. Water can then be reused.

ing water also passes through a heat exchanger or other apparatus where cooling occurs. Temperature increases from 6 to 20°F. Heated water is

263

then cooled in another exchanger using an external cooling medium, such as an air-cooled radiator.

Q   Is there a trend toward the use of circulating systems?

A   Since the once-through system calls for an almost unlimited supply of cooling water, today's trend is toward the use of circulating systems where the water is cooled for reuse. Cooling is brought about in an open system by actually evaporating a part of the cooling stream. In a closed system the cooling is usually done in an air-cooled radiator. A separate circulating or once-through setup can also be used.

Q   Where are open and closed systems used?

A   The open system is most often used for cooling water in chemical plants, oil refineries, and power-plant condensers. The closed system is usually limited to jobs where the quantity of cooling water isn't large. Typical applications include compressor cooling jackets and diesel engines.

Q   What causes scale formation?

A   The most common type of scale found in heat-exchange equipment is calcium carbonate, which is formed by the breakdown of calcium bicarbonate in the cooling water at high temperatures. Magnesium compounds and calcium sulfate are rarely the cause of scale in cooling systems. Scaling is affected by temperature, rate of heat transfer, calcium, magnesium, silica, sulfate, alkalinity concentrations, and the pH (see last question in this chapter for meaning of pH).

Q   How is the Langelier index used?

A   The Langelier index measures the tendency of calcium carbonate to precipitate from water under given conditions of calcium hardness, alkalinity, pH, temperature, and total dissolved solids. Charts, nomographs, and special slide rules can be used to calculate this index quickly for any water supply when the analysis is known. A positive index means the water has a tendency to deposit scale, while a negative index shows a tendency to dissolve scale and corrode. Note that the index does not measure the amount or rapidity of scaling. This depends only on the calcium carbonate or bicarbonate content. The index is decreased by reducing calcium hardness, calcium bicarbonate, pH, or a combination of all three. The total dissolved solids have a relatively minor effect on the index. Blowdown can be used to reduce the calcium and alkalinity in the circulating water, but it increases the make-up and is seldom used as the only means of control. Other treatment methods used to adjust the analysis include cold-lime softening, usually followed by an acid feed for the adjustment of pH and alkalinity, and partial zeolite softening with hard water bypass, also followed by an acid feed if needed. Some moderate-hardness high-alkalinity waters can be handled with an acid feed alone.

Q What is the "controlled scale" method?

A It is a method of keeping the index of the circulating water slightly positive, say in the range of 0.5 to 1.0 at the highest temperature in the system. This is usually enough to ensure the deposit of a thin impervious layer of calcium carbonate scale on the surfaces of the cooling-water system. Careful adjustment of pH, alkalinity, and calcium is needed to keep the desired index and protect the system against the ravages of excess scaling and corrosion.

Q What is the role of surface-active chemicals?

A Surface-active materials prevent crystal growth and therefore scale formation. They increase the solubility of scale-forming salts so precipitation doesn't occur when solubility limits are exceeded. Chemicals used for this purpose include special types of phosphates, tannins, lignins, etc. They can handle the job of scale prevention alone and are also helpful in broadening the range of Langelier index control in the controlled-scale treatment.

Q How do you control cooling-water system corrosion?

A Corrosion inhibitors such as chromates, phosphates, silicates, and alkalies are used. Anodic inhibitors of the chromate and phosphate type decrease metal attack. But if the concentrations are too high, pitting and tuberculation result. The use of chromate in combination with phosphate has been successful in controlling pitting, tuberculation, and other metal losses from corrosion; and dosages are at economical levels. Since oxygen and carbon dioxide are primary causes of corrosion, one of the more direct treatments is deaeration. With a vacuum deaerator oxygen can be reduced to as low as 0.1 parts per million (ppm), depending on the water temperature and equipment design.

Q What causes the formation of slime and algae?

A Algae growths are composed of millions of tiny plant cells, which multiply and produce large masses of plant material in a short time. Slime growths are a gelatinous mass of microorganisms, which cling tightly to secluded surfaces in the system, trapping organic and inorganic matter and debris along with scale-forming materials. Any appreciable build-up seriously interferes with heat-transfer efficiency.

Q What chemicals give effective treatment?

A Chemicals that prevent slime and algae troubles are those that exert a toxic action on the microorganisms. Chlorine is generally toxic for most bacteria, algae, and protozoa, but continuous application in a once through system is expensive. Intermittent treatment gives good results at a much more reasonable cost. Copper sulfate is toxic to simple algae but less toxic to slimes. Its use is limited because copper sulfate precipitates at a pH of 8.5 or higher. Phenolic and other quarternary ammonium compounds are also used. They have the advantage of resisting precipitation at a high pH and are not removed from the system by aeration in the cooling tower.

Q   Why is an intermittent feed best?

A   Intermittent large doses of a chemical, rather than a steady feed of a much smaller quantity, keep the microorganisms from getting used to the treatment and building up immunity to it. Intermittent dosage also avoids the danger of the development of a new strain of algicide- or slimicide-resistant microorganisms.

Q   How are accumulated slime and algae removed?

A   The best way to get rid of these growths is by mechanical cleaning. In the case of algae the algicide may only loosen the growth enough to set it free in the system to plug lines and damage pumps. After the system is completely cleaned, start out on the right foot with a good chemical treatment program to prevent future build-up of these growths.

Q   What causes cooling-tower wood destruction?

A   This destruction is usually caused by one or a combination of the following: (1) chemical attack commonly known as delignification, (2) mechanical rupture of wood cells from salt crystallization, and (3) biological attack causing wood decay.

Q   What is delignification, and when does it take place?

A   Delignification is the removal of tannins, lignins, and cellulose products from the lumber, leaving long stringy fibers with greatly reduced strength. It apparently occurs in the presence of high concentrations of sodium carbonate in the circulating water. Oxidizing agents, such as chlorine, speed up the process.

Q   Why are salt deposits harmful?

A   Intermittent splashing on parts of the cooling tower produces concentrated salt-saturated water in these areas, which is absorbed by the wood. Then crystals from beneath the surface of the wood mechanically disintegrate it. The result is a mass of soft, loose fibers susceptible to both chemical and biological attack.

Q   What is fungus attack, and how is it caused?

A   It is a form of biological attack that is caused by microorganisms that use the wood for food. These wood-destroying fungi reproduce by means of spores, which are normally airborne and alight on the surface of the wood. The attack is not confined to any particular section or make of cooling tower. It is most noticeable in the mist sections and takes place on both the surface and the interior of the wood.

Q   How do you combat fungus attack?

A   Proper control of the alkalinity, pH, and solids content of the circulating water will go a long way toward preventing chemical attack on the wood. It will not prevent biological attack but will decrease the chance of damage from fungus.

One popular preventive today is the use of properly impregnated cooling-tower lumber. Chromated copper arsenate salts used for this show much promise. A different approach is to try to destroy the fungi chemically in the circulating water by treating it with sodium penta-chlorophenate, zinc sulfate, and sodium chromate. Use intermittent dosing. It is often effective to spray a concentrated solution of toxicant directly on the cooling-tower wood in those sections most likely to be attacked by the fungus, such as tower mist areas.

Q   How is cooling-tower blowdown calculated?

A   The amount of blowdown usually depends on how far the circulating-water dissolved solids are allowed to build up by concentration. The number of concentrations N is one way of comparing dissolved solids in the make-up water with that of the circulating cooling-tower water. Since the chlorides remain soluble in concentrations, N is conveniently expressed as the ratio of chlorides in the circulating water to chlorides in

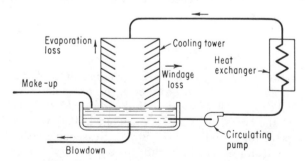

Fig. 18-2. Cooling-tower evaporation increases solids concentrations in recirculating water; blowdown and windage decrease them.

the make-up water. The system in Fig. 18-2 shows the basic items that are part of every cooling-tower system.

C = water circulating in cooling system, lb per hr

E = water vapor evaporated from tower, lb per hr
B = water lost by windage, plus that discharged to waste as cooling-tower blowdown.
MU = make-up water added to system, lb per hr
Cl_c = chloride content of circulating water, ppm
Cl_{mu} = chloride content of make-up water, ppm

Make-up water is added to compensate for water losses of evaporation, windage, and blowdown:

(1) $MU = E + B$

At equilibrium the amount of chlorides entering and leaving the system must be equal. The amount lost from the system in windage and system blowdown must equal the amount added in the make-up.

(2) $MU(Cl_{mu}) = B(Cl_c)$

The number of concentrations N by our definition is $Cl + Cl_{mu}$. Transposing (2) gives:

$$N = \frac{Cl_c}{Cl_{mu}} = \frac{MU}{B}$$

Substituting from Eq. (1),

(3) $N = \dfrac{E + B}{B} = \dfrac{E\% + B\%}{B\%}$

The amount of blowdown needed to limit concentrations in the circulating water is then easily found:

(4) $B\% = \dfrac{E\%}{(N - D)}$

The evaporation loss varies with the temperature drop of the circulating water. The evaporation of 1 lb of water in the tower absorbs about 1000 Btu. This will cool 100 lb of water 10°F. So, for each 10°F cooling, about 1 per cent of the circulating water is evaporated. The windage loss varies with the type of tower and local conditions. Reasonable estimates are 0.1 to 0.3 per cent for mechanical-draft towers, 0.3 to 1.0 per cent for atmospheric towers, and 1.0 to 5.0 per cent for systems using spray ponds.

BRINE SYSTEMS

Q   Where are brine coolers used?
A   Brine coolers are used with indirect refrigeration systems. Brine is

known as a secondary refrigerant. The cooler is usually the submerged type, similar to the shell-and-coil condenser. The cooler tank can be either cylindrical or rectangular, with an expansion coil placed inside. The tank usually is constructed of steel and is well insulated. The brine storage tank acts as a reservoir and is placed in the suction line to the brine circulating pump. The brine system is economical, safe, and ensures even temperatures in the compartments to be cooled.

Q Name the kinds of brine used.

A Brine may be made with common salt (sodium chloride). Another common brine is calcium chloride. The selection of the proper brine depends on the type of cooler and the required service. If ordinary pipe coils are to be used for cooling the brine and the temperatures to be maintained are high, sodium chloride is good. But if either the shell-and-tube or the double-pipe cooler is used with very low temperatures, calcium chloride is best.

Q What should the density of brine be?

A The density of the brine should be such that the freezing point is from 10 to 15°F below the temperature to which the brine is to be cooled in the brine cooler or ice tank. Then there will be no danger of the brine freezing. The brine temperature upon leaving the cooler should also be about 15°F below the temperature to be held in the coldest rooms.

EXAMPLE: With a room temperature of 24°F and a brine temperature of 9°F, the freezing point of brine is −5°F.

Q How can you detect ammonia leakage in brine?

A Ammonia in the brine can be detected by smell when more than a trace is present. Smaller amounts can be detected by heating 50 to 100 milliliter (ml) of brine in an erlenmeyer flask in which a moist strip of red litmus paper is hung. The litmus paper turns blue when a small amount of ammonia is present. Adding a little alkali (caustic soda) helps to free the ammonia more readily. The presence of an appreciable quantity of ammonia also increases the brine's pH. Litmus paper may not give reliable results when dipped into strong alkaline calcium chloride brine because this salt itself tends to turn the paper blue.

Q What are some qualities that a refrigeration brine should have?

A An efficient refrigeration brine must remain liquid at the lowest temperature at which it is to be used. Also, its corrosive effect on piping and cans must be held to a minimum. Freezing would form ice on evaporator coils, reducing their heat transfer, and in time would choke up restricted brine passages. Operation at low temperatures requires brine at higher specific gravity, which results in its having a lower specific heat. Corrosive brines cannot be tolerated because replacing metal cans and piping is costly. Because iron corrodes more quickly in brine of a low pH and zinc is damaged at relatively high pH values, it is very important to maintain the pH level at the best intermediate point, which is about 7.5 to 8.5. Refrigeration brine must be tested monthly for specific gravity, pH, free ammonia, and percentage of corrosion inhibitor.

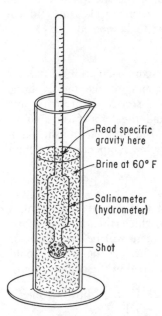

Read specific gravity here

Brine at 60° F

Salinometer (hydrometer)

Shot

Fig. 18-3. Salinometer is used for testing the specific gravity of brine.

Q Explain how to test brine for specific gravity.

A See Fig. 18-3. Pour a sample of brine (at 60°F) in a tall cylinder or salinometer jar. Float a hydrometer or salinometer in the liquid, and read the scale where it coincides with the brine level (not top of meniscus). Insert a thermometer to determine the brine temperature; if it is above or below 60°F, the specific gravity reading must be corrected by a factor given in brine tables. If the specific gravity of the brine is not within the desired range, make the necessary adjustments.

Q What does the pH reading of brine tell the operator? How would you measure the pH?

A The pH reading is the degree of the brine alkalinity or acidity. A pH of 7.0 represents neutrality. Values above this represent increasing alkalinity, and values below this represent increasing acidity. You can measure pH by either potentiometric or colorimetric methods. The colorimetric method compares the test sample against adjusted color standards. You can use a slide comparator-indicator solution combination that registers the brine's correct pH near the mid-point of comparator scale.

SUGGESTED READING

Reprints sold by *Power* magazine, costing $1.20 or less
Water Treatment, 56 pages
Corrosion, 32 pages

Books

Elonka, Steve, and Joseph F. Robinson: *Standard Plant Operator's Questions and Answers*, vols. I and II, McGraw-Hill Book Company, New York, 1959.

Elonka, Steve: *Plant Operators' Manual*, rev. ed., McGraw-Hill Book Company, New York, 1965.

Elonka, Steve, and Alonzo R. Parsons: *Standard Instrumentation Questions and Answers*, vols. I and II, McGraw-Hill Book Company, New York, 1962.

19

SAFETY

Q Describe briefly the refrigeration codes.

A Refrigerating equipment should be designed, installed, and repaired according to the ASA Code B9.1 (Safety Code for Mechanical Refrigeration, American Standards Association); also according to the code and regulations of the National Board of Fire Underwriters (NBFU), Standards for the Unit Refrigeration Systems and Standards for Air-Conditioning and Commercial Refrigeration Equipment; and according to all the local municipal and state codes.

These codes cover refrigeration plant location in respect to flammability and toxicity, type of occupancy of the building, foundations, electrical equipment and wiring, machine rooms, ventilation, pipes and fittings, strength and test pressures, safety devices, gages, discharge means, markings, adding or withdrawing of refrigerants, testing for leaks, gas masks, maintenance, and periodic inspection. Codes also provide tables showing permissible pressures, quantities of various types of refrigerants allowed, sizes of air ducts and openings, line sizes, and sizes of relief devices.

If there is no state or local code, or if the local codes do not cover specific installations, follow the B9.1 code. Consult it when the equipment is installed and tested. The suppliers should be asked to furnish evidence of compliance with the requirements of the local codes.

Electrical equipment, wiring, and motors should comply with the national code (NBFU) and ASA. All persons responsible for installation, testing, operation, maintenance, and repair of refrigerating equipment should know the provisions of all codes which apply. After major repairs the system should be tested in compliance with the codes as if it were a new installation.

Q Name the major hazards encountered with refrigeration.

A The main hazards are (1) explosions, (2) fire, and (3) toxic effects from the gases used. The explosion hazard tends to increase as more refrigerant is used. If gas escapes, it may damage goods in cold storage. If

272

gas is toxic, it may cause serious personal injury. If it is flammable, it may form an explosive concentration.

Q   What are some of the main causes of explosions in the crankcase of a refrigeration compressor?

A   Two main causes of explosions in the crankcase of a compressor are shots of liquid impurities in the gas and leakage past the piston rings and packing. This leakage allows oil vapors to mix in the crankcase with the refrigerating gas vapors at high temperatures, thus creating a flammable mixture.

Q   What safety feature should a gage glass installed in a refrigeration system have?

A   A gage glass in a refrigeration system should have a device which will close automatically if the glass breaks. The glass also should have a guard to protect it against external blows.

Q   What safety rule must every maintenance and repair man observe before working on machinery?

A   Before starting work on any engine or motor, line shaft, or other power-transmission equipment or power-driven machine, make sure that it cannot be set in motion without your permission.

Q   Explain four ways to prevent machinery from being set in motion.

A   (1) Place your own padlock on the control switch, lever, or valve, and keep the key in your pocket.

REMEMBER: Even if someone has locked the controls before you, you are not protected unless you put your own padlock on it.

(2) Place a "men at work" sign on the control, and padlock or block the mechanism in some effective way. Never allow someone else to do this for you. (3) Be sure that in removing your padlock and sign you don't expose someone else to danger. (4) If a duplicate key to your padlock is lost, get a new padlock.

Q   What precaution would you take before opening a system for repairs?

A   No part of a system should be opened for repairs until all pressure has been relieved from that section and all refrigerant gas pumped out of it. The section to be repaired should be pumped down to a vacuum of 10 to 15 in. if possible, or to at least atmospheric pressure, and held for a minimum of 10 min. To assure that all refrigerant has been removed, repeat the pumping down at least three times. Pour water on the piping as a further test. If frost appears, there may still be too much refrigerant inside to open the section safely.

Q   How does freezing cause explosions, and how would you protect against it?

A   Because a frozen part often cannot function, dangerous pressures are built up in the system as the compressor continues to operate. Then an explosion results. To prevent overpressure from this cause, install a pressure differential switch across the pump or a water-pressure failure switch at the condenser.

Q   Are fire extinguishers needed to protect refrigeration systems?

A   Yes. Fire extinguishers of approved types should be kept near the equipment and doors or at other accessible locations in the machine room. The fire extinguishers should be plainly visible and well marked. Operators should know their location and be instructed in their use. Vaporizing liquid and carbon dioxide extinguishers are suitable for electrical and refrigeration fires. The carbon dioxide type is preferable because it is not as toxic as the vaporizing liquid on a hot fire. For emergencies it is good practice to have two or more doors to all refrigeration machine rooms.

Q   Can you use a gas mask of the canister type for all types of refrigerants?

A   No. The type of canister should be approved by the U.S. Bureau of Mines for the refrigerant being used in the system. Gas masks of the canister type do not give suitable protection against either carbon dioxide gas or oxygen deficiency. An oxygen-supplied gas mask is the only type which protects in such cases.

Q   When should gas masks be worn?

A   Gas masks should be used if even small concentrations of ammonia, sulfur dioxide, or methyl chloride are present in the air. They should also be used when the escape of some of the less toxic refrigerant gases seems likely or if the presence of products of decomposition or combustion is suspected.

Q   Is it dangerous to smoke while drawing oil from an oil trap?

A   Yes. Oil vapor and ammonia gas explode from an open light. In one plant when a gasket blew out of a compressor head the operator pulled the switch because the plant was filled with ammonia fumes. The switch arced, and an explosion wrecked the plant. Freon decomposes in a fire to form extremely toxic gases.

Q   Should you test your system from time to time?

A   City codes specify various test pressures for the low and high side with various refrigerants. A common practice is to test the ammonia high-pressure side with air of 250 to 300 psi and the low-pressure side with air of 150 psi. When pumping air into a large system, don't use an ammonia compressor with a bore over 4 in. The speed on larger units cannot be controlled; they heat rapidly under high pressure. The best setup is a 4 × 4-in. compressor belt driven at 200 rpm. Use only an air-compressor oil in the refrigeration compressor if using it as a compressor. Connect the compressor discharge line to some point, such as the receiver, where air, moisture, and oil can later be blown out.

Q Are refrigerants toxic?

A In a strict sense all refrigerants except air can cause suffocation by oxygen deficiency. The term "toxic" is usually applied to refrigerants which are actually injurious to human beings. Some cause death or serious injury even though mixed in small doses with air. In these cases enough oxygen cannot be obtained from the air-refrigerant vapor mixture. A mixture of toxic vapors and air may attack the membranes of the lungs and be carried to other parts of the body. So, know your refrigerants and what to do in case of an accident.

Q How would you enter a room having a bad ammonia leak?

A Before entering a room where there is an ammonia leak, put on canister-type gas mask approved for ammonia use by the U.S. Bureau of Mines. Then spray the room with water because it absorbs ammonia quickly and washes it to the floor. Use rubber gloves to protect your hands. Don't use an all-purpose gas mask if the air contains more than 3 per cent ammonia. You will know when that point is reached because ammonia will start to penetrate the mask's filter.

Q What are some of the safety precautions that should be observed in a plant using ammonia?

A The Manufacturing Chemists Association and the Compressed Gas Association have both published extensive recommendations and suggestions for safety and first aid in connection with ammonia. Following are some of the points covered: While ammonia is not a poisonous gas, it does severely irritate the mucous membranes of the eyes, nose, throat, and lungs. A very small concentration is easily detected by its sharp pungent odor. See Fig. 19-1 for exposure limits and concentrations. Liquid ammonia should never come in contact with the skin because it freezes tissue, subjecting it to caustic action. Symptoms of such action are similar to symptoms of a burn. Employees handling ammonia should understand the possible hazards and should be instructed and trained in the following: (1) Location of gas masks and other protective equipment. (2) Location of safety showers, bubbler drinking fountains, water hoses, exits, and first-aid equipment. An adequate water supply is ex-

Gaseous, ppm	Concentration, per cent	Effects on unprotected worker	Exposure period
50	0.006	Least detectable odor	Permissible for 8-hr working
100	0.0125	No adverse effects for un- protected worker	exposure
400	0.05	Causes irritation of throat	Ordinarily no serious results
700	0.1	Causes irritation of eyes	following infrequent short exposures (less than 1 hr)
1,720	0.22	Causes convulsive coughing	No exposure permissible— may be fatal after short exposure (less than ½ hr)
5,000–10,000	...	Causes respiratory spasm, strangulation, asphyxia	No exposure permissible (rapidly fatal)

Fig. 19-1. Exposure limits and concentrations of ammonia.

tremely important because quick and thorough washing is essential when ammonia comes in contact with eyes or skin. (3) Proper use of gas masks and other protective equipment. (4) The urgency of immediately reporting any unusual odor of ammonia. (5) Proper conduct in case of an emergency. (6) Methods of properly handling ammonia containers and approved procedures for cleaning pipes and other equipment.

Q   What action should be taken if someone is burned or overcome by ammonia?

A   As with most serious accidents, it is a good procedure to call a doctor at once. The following general measures will be helpful:

Ammonia gas. (1) Give patient fresh air, preferably in a warm room where the temperature is about 70°F. (2) Lay the patient down with head and shoulders slightly elevated, and keep him warm with blankets or other clothing. (3) When breathing is weak, administer oxygen. (4) If breathing has stopped, immediately start artificial respiration.

Liquid ammonia burns. (1) *Eyes:* Wash eyes with water immediately. Use a drinking fountain or hose or plunge head in a container of water, repeatedly opening and closing the eyes. Wash continuously for 15 min. A mild (2 to 5 per cent) boric acid solution may be used if available. (2) *Skin:* Immediately remove any clothing saturated with ammonia. Quickly and thoroughly wash the burned area with cool water. After a thorough washing apply lemon juice, vinegar, or 2 per cent solution of acetic acid. Never use salves or ointments. Ice applied directly to burned areas helps relieve pain. (3) *Nose and throat:* Inhale 2 per cent boric acid solution up the nose and rinse the mouth thoroughly. If the patient can swallow, have him drink large quantities of ½ per cent citric acid solution or lemonade.

Q Is ammonia gas dangerous when in contact with flame?

A Ammonia gas itself is not explosive and doesn't burn easily. At high temperatures, it burns with a greenish-yellow flame. It decomposes at about 800°F into its constituent parts of nitrogen and hydrogen. Hydrogen burns, and when mixed with air in the right proportions, it is explosive.

Q What should every operator know about the dangers of carbon dioxide?

A When carbon dioxide leaks, it mixes with air. There is nothing to worry about if the leak is small. But when air is mixed with more than 5 per cent carbon dioxide, you get headaches and feel drowsy; larger doses will cause unconsciousness. When the air contains about 9 per cent carbon dioxide, you suffocate. So, if the leak is large, use an oxygen-breathing apparatus or an air-line mask when fixing it. Don't use an all-purpose gas mask, because it relies on the atmospheric condition in the room. Human beings suffocate with less than 7 per cent oxygen in the air. The flame of a safety lamp goes out when there is 16 per cent oxygen in the air. If this flame won't burn, don't use an all-purpose gas mask.

Q Are Freon-type refrigerants (halocarbons) dangerous?

A Halocarbons form a nonpoisonous and nonirritating gas; but a large leak (over 20 per cent concentration with air) causes unconsciousness. The refrigerant itself won't burn, but in the presence of a flame or hot surfaces above 1000°F it decomposes into toxic products that are extremely irritating and poisonous. So take all precautions when welding. Unlike ammonia, a halocarbon refrigerant does not form explosive mixtures when it breaks down into its constituent parts.

Q Is it safe to use carbon tetrachloride as a cleaning solvent?

A Carbon tetrachloride is often used as a cleaner, but it is a killer. It evaporates rapidly and has a very toxic vapor. In a confined space with no direct ventilation the vapor quickly overcomes a worker. It can also harm the skin and body organs.

To avoid the flammable or explosive hazard of low flash-point solvents and kerosene, carbon tetrachloride is often mixed with them. While this reduces the fire and explosion hazard, it does not remove the toxic vapor hazard. It is difficult to ventilate confined spaces around a compressor in order to eliminate these toxic vapors unless you use a forced-circulation blower.

Q Why shouldn't gasoline be used as a cleaning solvent?

A Gasoline is dangerous because its vapors not only are explosive, but also have a toxic effect on the respiratory system. Gasoline also burns the hands and skin, causing infection or dermatitis. The chemical action of gasoline dries out the skin's natural protective oils. The most dangerous hazard is the explosion of gasoline vapors. A hazard also exists when other low-flash-point solvents, such as naphtha or benzol, are used. When gasoline is vaporized in air it has three times the explosive energy of TNT. To make this powerful explosive the vapor must be mixed with the air. It explodes at different mixtures; the most dangerous is a 6 per cent gasoline mixture with 94 per cent air by volume. Thus a small residual coating of gasoline left in the crankcase after cleaning can generate enough explosive vapor to blow up the machine.

Q What is a safe cleaning solvent?
A A good, safe cleaning solvent is 25 per cent methylene chloride, 70 per cent stoddard solvent, and 5 per cent perchlorethylene.

CAUTION: Avoid contact with skin or breathing the fumes.

Q Dry ice (solid carbon dioxide) is used around plants for making shrink fits; it has many other applications. What are some safety precautions to take when using it?
A Because the temperature of dry ice is $-109°F$, it can cause frostbite, similar to a severe burn. If it is kept in a tight container, the pressure of dry-ice gas can cause an explosion, so take the following precautions: Wear gloves when handling either dry ice or metal that has been in contact with it. Never put dry ice in a drink; a small piece can cause frostbite. Never place dry ice in a tight container other than one that is specifically designed for it. When sawing dry ice, see that the guards are in place and use a wooden strip to push the dry ice against the saw blade. In a small enclosed space gas from the dry ice can cause suffocation; if breathing quickens, dizziness occurs, or your ears start to ring, get to fresh air at once. Learn to use artificial respiration.

Q Explain why low voltage is a killer.
A You cannot feel an electrical current of 1 milliampere (ma) or less. From 1 to 8 ma the shock isn't painful and muscular control is OK. From 8 to 15 ma the shock is painful, but muscular control isn't lost and you can let go. From 15 to 20 ma the shock is painful, control of the muscles is lost, and you cannot let go. From 20 to 50 ma there is painful, severe muscular contraction and breathing becomes difficult. A 50- to 100-ma charge often causes ventricular fibrillation (a heart condition resulting in instant death). A 200-ma charge, or more, causes severe burns and muscular contraction. Chest muscles clamp the heart and stop it for the duration of the shock. One hundred and ten volts alternating current is a killer.

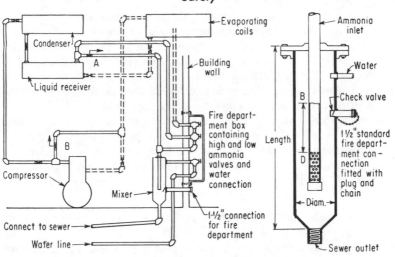

			Mixer							Hand valves		Safety valves	
Rated tonnage	H.p. of motor	Charge of ammonia, lbs.	Diameter inches	Length feet	Sewer outlet	Number of 3/32" holes	Ammonia inlet	Water inlet	No. of inches B to D	Low side	High side	On (A) receiver	On (B) compressor
Up to 1/2	1	25	3	4	2	120	3/8	1 1/2	12	3/8	3/8	1/4	1/4
1	2	50	4	4	2	180	3/8	1 1/2	12	3/8	3/8	3/8	1/4
2	3	75	5	4	2	200	1/2	1 1/2	12	3/8	3/8	3/8	3/8
3	5	100	5	4	2	200	3/4	1 1/2	12	1/2	1/2	3/8	3/8
5	7 1/2	200	5	4	2	240	3/4	1 1/2	12	1/2	1/2	1/2	3/8
10	15	300	5	5	2 1/2	280	1	1 1/2	18	3/4	3/4	1/2	1/2
15	25	450	5	5	2 1/2	280	1	1 1/2	18	3/4	3/4	1/2	1/2
25	40	600	6	5	2 1/2	300	1 1/4	1 1/2	18	1	1	3/4	1/2
30	50	750	6	5	2 1/2	360	1 1/4	1 1/2	18	1	1	3/4	1/2
50	75	1200	6	6	3	420	1 1/4	2	20	1	1	3/4	3/4
75	125	1600	7	6	3	540	1 1/2	2	20	1 1/4	1 1/4	1	3/4
100	150	2000	7	6	3	620	1 1/2	2	20	1 1/4	1 1/4	1	1
150	250	3000	8	6	3	900	2	2	20	1 1/4	1 1/4	1 1/4	1
200	300	3500	8	6	3	1000	2	2	20	1 1/2	1 1/2	1 1/4	1 1/4
300	450	5000	10	6	4	1400	2 1/2	2	24	2	2	2	2
600	900	10000	14	6	4	2500	3	3	24	2	2	3	2 1/2

Fig. 19-2. Safety connections are used to dump ammonia to the sewer.

Q How would you dump ammonia to a sewer in an emergency?

A See Fig. 19-2. Escaping ammonia can drive rescue squads and fire-men away in an emergency. The sketch shows how to construct and install a disposal unit to dump ammonia to a sewer if a sudden leak or fire develops. When planning the layout, put all shutoff valves at the mixer. A closed valve at the vessel to be dumped destroys all the value of the disposer during an emergency. Mount the mixer on the outer wall so that the shutoff valve can be operated from the outside. Use the table to find the size of the mixer that fits the capacity of your plant. But before going to the expense of installing such a system, consult local authorities about dumping ammonia to a sewer. Check your local code on this and all other requirements. Your insurance company is also interested.

Q How would you prevent slugs of ammonia from returning to the compressor in a large, overworked system?

A See Fig. 19-3. The common practice of overfeeding liquid ammonia to the evaporator coils creates the danger of liquid-ammonia slugs going

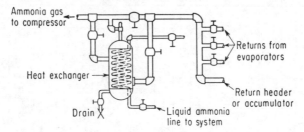

Fig. 19-3. Hookup prevents slugs of ammonia from returning to the compressor.

to the compressor. In some cases the compressor head blows off, killing or maiming someone. Also, ammonia that isn't vaporized does no cooling and is wasted. The hookup shown in the sketch not only overcomes the danger of slugs in plants working at peak loads, but also gets full value from the ammonia.

This hookup is a common heat exchanger installed on the low side of the line. All return lines from the expansion coils enter into a common header or accumulator. Instead of the suction line from the compressor being hooked directly to the accumulator, it has bypass valves and is hooked to the heat exchanger. The liquid-ammonia line is piped through the heating coil in the heat exchanger. Any liquid ammonia carrying over from the evaporator picks up enough heat from the liquid-ammonia coil to vaporize before passing out of the exchanger; therefore piping should be insulated.

Q Is oil in a refrigeration system dangerous?
A Oil has a bad effect on a system—this is why an oil separator is used. The separator should be blown down regularly so that any entrained oil may be drained out. Always lead the drain from the oil separator to a tank or covered vessel containing water. Operators have been badly burned by ammonia flowing from an open oil-separator drain. A sudden spurt of oil and ammonia splashing into water can cause serious burns.

Q Would you use oxygen to test a refrigeration system for leaks?
A No. Oxygen mixed with oil is very explosive. It will explode without a spark.

Q What should every operator of refrigeration equipment know about the canister-type gas mask?
A See Fig. 19-4. Contaminated air is purified by chemicals in the canister, but no one chemical will remove all gaseous contaminants.

Fig. 19-4. Proper use of gas mask must be thoroughly understood by every operator.

Canister gas masks with a full facepiece are for emergency protection in atmospheres dangerous to life. Their effectiveness is limited to concentrations of 2 per cent by volume for all contaminants except ammonia, which is 3 per cent.

Persons who will be using a mask should be checked for heart and lung conditions. Make sure that the mask fits: (1) Remove the plug from the bottom of the canister and put on the headpiece. (2) Adjust the head straps until the mask fits closely and comfortably. (3) Avoid kinks in the tubes or hose. (4) Breathe naturally and observe the amount of resistance to inhalation. (5) Grasp the hose tightly and close off the air intake. Breathe deeply until the facepiece collapses. If it is correctly adjusted, the facepiece will remain collapsed until the intake is opened. With the mask on enter the contaminated area cautiously. If the mask leaks or the canister is exhausted, you'll know by odor; taste; or irritation of eyes, nose, or throat; and you should return immediately to fresh air. If the canister is used up, remove it and replace it with a new one. If the respirator is worn in a gas or vapor having no warning properties, such as carbon monoxide, use a fresh canister each time, unless the mask has a timer. Even though some masks do not have timers, a time indicator is important.

Train each new employee to wear the mask long enough to become accustomed to its breathing resistance. Remove the seal used to prevent air entering each time a new canister is installed in the mask.

Q   How many helmets and gas masks are required in a refrigerating machinery room? Where should they be located?

A   Class A systems with over 1,000 lb of refrigerant should have two masks in the machinery room near the main exit and one or two masks at convenient points throughout the plant. All codes approve masks approved by U.S. Bureau of Mines for the refrigerant in use.

Q   Are open-flame lights permitted in the machinery rooms?

A   Arc lights or other open-flame lights are not permitted.

Q   What are the general requirements for ventilation in refrigerating machinery rooms?

A   Requirements vary with the amount of refrigerant in the system, the type of refrigerant, and whether the ventilation is natural circulation or forced-draft. Thus, a Class A system with over 1,000 lb of refrigerant needs (1) an exhaust fan with a capacity of at least 2,000 cfm, or (2) a window area to outside air of 25 sq ft if on opposite walls, or (3) a window area of 60 sq ft if windows open on one side of machinery room. See your local code.

Q   If you are testing a new system using anhydrous ammonia as a refrigerant, what test pressures are required?

A   Most codes require pressures of 300 psig on the high-pressure side and 150 psig on the low-pressure side. Some codes restrict the use of air for test purposes.

   IMPORTANT: Take great care in pumping air pressures on systems for testing.

It is not advisable to use large ammonia compressors for pressures in excess of 150 psi. Above this point use a small air compressor designed for high-pressure work and lubricated with air-compressor oil. The best procedure is to open all the valves on the low-pressure side of the system. This permits the maximum allowable pressure on brine coolers and accumulators to equalize with that of the high-pressure side. The entire system may then be tested at 150 psi. The next step is to close off the low-pressure system and blow out the air pressure, leaving a valve open at some point to the atmosphere. This prevents the possibility of placing a higher pressure than would be safe on the low-pressure side of system while testing the high-pressure side. The high-pressure side is then pumped to the required pressure and tested for air leaks. Soapy water applied with a soft paintbrush at all joints is a simple way to locate small leaks. After checking over the entire system by this method, blow out the air pressure, close off the atmospheric connections to the low side, and see that the expansion valves are open. Then pump a vacuum on the system, exhausting all the air possible. Large compressors may be used for this purpose. The system is now ready for charging. It is good policy to charge in only a small amount of ammonia, enough to raise the gas pressure to 75 to 100 psi, and then again completely test the system for small gas leaks.

CHECKING OUT NEW REFRIGERATION SYSTEMS FOR SAFETY

Q List the check-outs to be made on a new system.
A Controls, interlocked system, regulating valves, expansion valve, thermostats, lube oil, piping, start-up checks, running check, and final check are recommended by the Hartford Steam Boiler Inspection & Insurance Co.

Q Explain the check-out of controls.
A Controls vary with the job type, but most air-conditioning and similar systems using reciprocating compressors have high- and low-pressure cutouts. This is called dual-pressure control. Set the low-pressure cutout so it starts the compressor when suction pressure reaches the value corresponding to the suction temperature desired. Set the high-pressure cutout also. Don't touch the pressure controls unless you have the manufacturer's instructions for them, or you may run into trouble.

Be sure to adjust differential points as low as you can without causing the compressor to start and stop too frequently (known as short-cycling).

Q How are interlocked systems checked out?
A Interlocked systems are popular for air conditioning and for jobs where it is necessary to prevent units from running in the wrong sequence. Interlocks are used to prevent compressor operation when (1) the con-

denser is out of service, (2) air-conditioning fans are not running, (3) a liquid solenoid valve is closed, etc.

To see that each interlock functions properly, you must set up conditions that would cause it to stop the compressor or prevent it from starting, depending on how the unit is to be locked out.

Q   How are regulating valves checked out?

A   Apply head pressure to the condenser and check the flow through the cooling-water regulating valve. Adjust the valve to give the right flow. If you cannot adjust it to give the desired flow, check the inlet pressure. Most water-regulating valves have an upper limit for inlet pressure, and if city water pressure is greater, a pressure-reducing valve must be used in the line.

Q   Explain how expansion valves are checked.

A   If these valves have been preset at the factory, it is best to operate the system to check refrigerant feed to the evaporator. Factory-set valves don't need adjustment often. But if the valve isn't preset, check the superheat adjustment by reading the refrigerant temperature in the suction line at the thermal bulb.

The best way to measure gas temperature is to install a thermometer well in the suction line close to the thermal bulb. If the suction tubing is too small for a well, clamp a thermometer to the outside of the tube. To ensure an accurate temperature reading, get as good a bond as possible between the thermometer bulb and the tube. Make a final adjustment of the expansion valve when the evaporator is running at or near the needed temperature.

Q   Should the thermostat be set, or checked?

A   A thermostat is set to keep a room at the desired temperature. If unwarranted resetting is to be prevented, protect the thermostat with a suitable guard.

Q   What check-outs should be made of lube oil?

A   Small compressors that are charged with refrigerant at the factory are also filled with the correct amount of oil. But before starting the unit, check the oil level to see that it is correct.

A large compressor that must be charged with refrigerant in the field must also be filled with oil before the unit is started. Use only the grade recommended by the manufacturer, keeping it clean and moisture-free by storing it in closed containers. Add oil either by pouring it into the crankcase or drawing it in, using the vacuum created by the compressor. See the manufacturer's instructions for the best procedure.

Q   What check-outs should new piping receive?

A   Never make the mistake of leaving rags, tools, or anything else in the piping. Check all lines and valves before charging the system.

See that gages are connected to the suction and discharge lines and to lube-oil piping. Open gage cocks; a closed cock can lead to trouble during start-up, resulting in damage to the system. Then connect all thermometers needed.

Q Explain how to start up a new refrigeration system, or one that has undergone extensive repairs.

A Open the condensing-water control valve and check the discharge of water from the condenser. Inspect the compressor to see that it is free to turn over and that no tools or other equipment are near the flywheel and drive. Check the oil level. Open suction and discharge valves in the refrigerant lines.

Bar the compressor through several revolutions by hand so that you are sure there are no obstructions to its free movement. If bypass valves must be open for correct starting, open them according to the manufacturer's instructions. Start and stop the compressor several times, allowing the unit to run for about 10 sec each time. Check system pressure gages.

Q Describe a running check of a new system.

A Operate the system for 72 hr, but keep a service man on the job for at least 48 hr. Watch compressor oil level closely at all times while the system is running. Take voltage and amperage readings on all motors at the start of the running check. This prevents damage from overheating. See that these readings agree with nameplate values.

Once the system carries load, examine all controls, resetting them if conditions are not right. Inspect piping for leaks, using a gas detector. Read and record operating temperatures and pressures. See that suction superheat is correct.

Q Explain the final check of a new or repaired system.

A After the system has run 72 hr it should be shut down and carefully inspected. A run of this duration gives the system's units time to run in, and any major faults should have shown up.

Check oil level, refrigerant flow, expansion-valve superheat setting, evaporator pressure regulator, head and suction pressure, shaft seal, alignment, pipeline strainers, and air-handling units.

Once these items have been checked and found OK, you are ready to run the system for a full week. Most compressor manufacturers recommend an additional inspection at the end of this period. See their instructions for the best procedure.

Hints given here are only general. Each make of compressor is a little different and you will have to follow slightly different procedures.

Q What precautions must one take when handling refrigerant cylinders, and why?

A   Only wrenches and other tools provided by the cylinder manufacturer should be used in opening or closing cylinder valves. When heat is applied to a cylinder while charging, never let the temperature go over 150°F. Fusible plugs (if mounted in the cylinder) usually melt at 150 to 175°F, or less at times. If cylinders are stored in an upright position, foreign material stays in the bottom away from the discharge line. Jarring or striking a charged cylinder may cause an explosion, so handle with *care*. Once charging is complete, immediately remove the cylinder. Make sure cylinders handled have built-in relief valves. Keep cylinders, whether full or empty, away from heat, and away from ice or snow in winter.

SUGGESTED READING

Reprints sold by *Power* magazine, costing $1.20 or less
    Refrigeration, 20 pages
    Power Handbook, 64 pages

Books
    Elonka, Steve, and Joseph F. Robinson: *Standard Plant Operator's Questions and Answers*, vols. I and II, McGraw-Hill Book Company, New York, 1959.
    Elonka, Steve: *Plant Operators' Manual*, rev. ed., McGraw-Hill Book Company, New York, 1965.
    ASHRAE Data Book, American Society of Heating, Refrigerating and Air-Conditioning Engineers.

20

CRYOGENICS

Q   What is cryogenics?

A   Cryogenics is a new field of engineering associated with the development, production, and use of equipment in the range of −300°F to absolute zero (see Fig. 20-1). It includes investigations of the properties

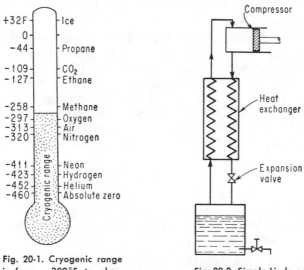

Fig. 20-1. Cryogenic range is from −300°F to absolute zero.

Fig. 20-2. Simple Linde or Hampton system.

of materials at very low temperatures, liquefaction of gases, and development of all types of equipment and instruments for such work. The name itself comes from the Greek *kryos*, meaning icy cold, and from *genic*, meaning to become or to produce.

Q   What are the two major classifications of industrial applications of cryogenics?

A   The first and still greatest application of the cryogenic field is the bulk production of liquefied gases. The second use is the application of very low temperatures to certain processes.

Q   What gases are liquefied, and what are the major uses of each?

A   Present cryogenic plants are designed for the liquefaction and separation of ordinary atmospheric air into its components of oxygen, nitrogen, argon, neon, helium, krypton, and xenon. Hydrogen, ethylene, methane, and other gases are liquefied from petroleum gas streams. New processes in the steel industry absorb 50 per cent of liquid oxygen production; the chemical industry uses 25 per cent; missiles and rockets use 20 per cent; the remainder is used for welding, medical therapy, and miscellaneous uses. Liquid nitrogen is used in metal treating, food freezing, and food transportation. Petroleum gases are liquefied for ease of transportation. Helium finds great use in shielded arc welding, rocketry, weather balloons, and low-temperature research. Liquid helium is also used in maintaining near absolute zero temperatures in infrared heat detectors.

Q   How is air liquefied?

A   The simple Linde or Hampton system is often used where small quantities of liquid air are required (see Fig. 20-2). This system consists of three minimum parts: a compressor, a heat exchanger, and an expansion valve. The air leaves the compressor at about 3,000 psig, is cooled in passing through the heat exchanger, and expanded adiabatically to atmospheric pressure. About 10 per cent of the air leaving the expansion valve liquefies; the remaining unliquefied air passes through the cold side of the heat exchanger, cooling the high-pressure air leaving the compressor.

Q   What is the Joule-Thomson effect?

A   The Joule-Thomson effect is the change in temperature resulting from a gas expanding from one constant pressure to a lower constant pressure, with no heat transfer and no external work other than to maintain the constant pressures.

Q   What is an expansion valve?

A   An expansion valve is an adjustable orifice through which a Joule-Thomson expansion occurs.

Q   What is an expansion engine?

A   An expansion engine is a machine used to extract work (or energy) from a high-pressure gas stream by causing the gas to drive a piston or turbine wheel as it expands to a lower-pressure level. As the energy is removed by the engine, the temperature of the gas decreases. The energy taken by the expansion engine is returned to the system.

Q What is the Claude system?

A The Claude system is a combination of both the Joule-Thomson effect and the expansion engine (see Fig. 20-3). Air is compressed to about 600 psig, passes through the first heat exchanger, and separates into two streams. About 20 per cent passes through the heat exchangers and is expanded through the expansion valve to atmospheric pressure. Part of the stream is liquefied, and the remaining portion returns to the compressor through the heat exchangers.

Q What is the cascade system?

A The cascade system in Fig. 20-4 uses ammonia, ethylene, methane, and nitrogen as refrigerants. These all have progressively lower boiling points. The evaporator of one refrigerant is used to condense the discharge from the next-lower-temperature compressor. The total system horsepower input per unit of nitrogen produced is lower than that in other systems.

Q How are the component gases separated from a stream of liquid air?

A As indicated in Fig. 20-1, oxygen, nitrogen, neon, etc., all have different boiling points. When the air-stream

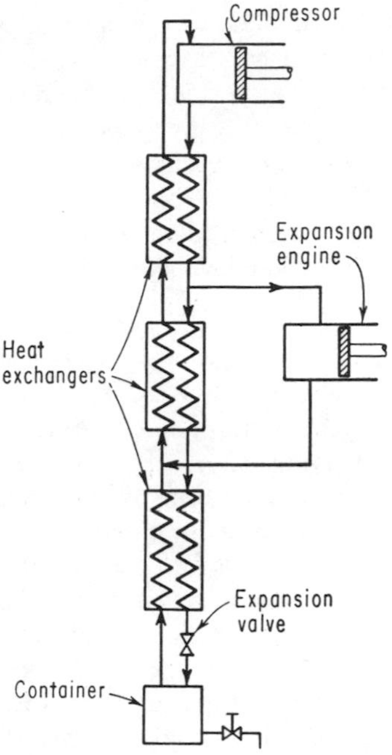

Fig. 20-3. Claude system combines Joule-Thomson effect and expansion engine.

temperature rises, each element boils at its own separate boiling point. By controlling temperatures through a fractionation column, similar to the bubble column in an absorption system, gases of a high purity are removed at different levels.

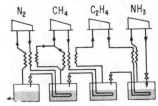

Fig. 20-4. Cascade system uses various refrigerants.

Q Do the low temperatures used in cryogenic plants have any effect on the materials used?

A Yes. The physical properties of materials change radically with a reduction in temperature. For instance, ordinary carbon steel becomes very brittle. Most copper alloys, stainless steels, and aluminum increase in tensile strength. Thermal conductivities of most pure metals may be several times greater than at normal temperatures.

Q   What types of thermometers or temperature-indicating devices are used for these low temperatures?
A   The gas thermometer is the most accurate means of determining true thermodynamic temperature according to the definitions of gas laws, since the pressure is proportional to the absolute temperature. Thermocouples and resistance thermometers are used extensively, since they can be located in inaccessible locations and are quite accurate. Liquid-in-glass thermometers filled with special fluids are sometimes used for tempera tures as low as about −300°F.

Q   Compare refrigeration and liquefaction machines.
A   The cryogenics industry deals with systems that liquefy gases either for consumption or, in the case of closed cycles, for use as a refrigerating medium. There are three basic methods of generating cold and liquefying gases. (1) Compression of a commercial refrigerant, followed by its evaporation, is the principle used in many large plants. The heat of evaporation represents the amount of refrigeration available. But this method is not an economical way to produce temperatures below −60°F in a single-stage machine. However, successive steps can be used, in cascade, to reach −300°F, a temperature low enough to liquefy oxygen.

(2) Use of the Joule-Thomson effect (discussed previously) is a more effective method used in the cryogenics industry. Here, a gas is compressed, then made to pass through a valve, where it is suddenly expanded. The resulting drop in pressure also brings about a substantial drop in temperature. Today's practice is to combine the Joule-Thomson valve with either the evaporation method or a cycle involving an expansion engine for more efficient results. However, each gas has a definite inversion temperature above which the gas is not cooled, and may even increase in temperature (for example, −109°F for hydrogen, −380°F for helium). Thus, it is customary to precool the gases to a level well below the inversion temperature by thermal exchange with another gas that is easier to liquefy.

(3) Expansion of the compressed gas in a mechanical expander is the third method; it gives a greater temperature reduction than the Joule-Thomson effect. The gas is compressed to pressures up to 3,000 psi, made to drive an expander (turbine), and then released at pressures (exhausted) that can be as low as 20 psi. A significant portion of the energy released during expansion can, of course, be recovered. The recovered portion can supply part of the energy needed for compression, or it can drive another machine. This is known as the closed cycle. While a somewhat large compressor is required for the closed cycle, increasing

the flow of a rotary compressor by as much as one-third does not involve much additional cost.

The expander may be of the reciprocating or rotary type. Where the volume of gas is relatively small (up to 1,000 cfm per cylinder) and where a high pressure is desired, reciprocating engines are used. But when large volumes of gas are handled (up to about 1.5 million cu ft per hr), expansion turbines of the radial or axial type are preferred. Fig. 20-5 is a schematic of the low-temperature liquefaction system used to produce liquid helium at −452°F.

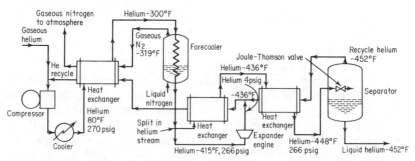

Fig. 20-5. Low-temperature liquefaction shows importance of coolers and heat exchangers in producing liquid helium at −452°F.

SUGGESTED READING

ASHRAE Applications Data Book 1971, American Society of Heating, Refrigerating and Air-Conditioning Engineers.

21

MANAGEMENT OF
REFRIGERATION SYSTEMS

Q. What are the requirements for efficient and safe operation of air-conditioning and refrigeration machinery?

A The first requirement is qualified personnel and knowledgeable owners. Operators should know the basics of the subject; they should know what takes place inside the equipment when a button is pushed, a valve opened, or a switch closed. The only way to know your plant is to trace out each line (piping, cables) and make a sketch. Where qualified personnel are not available for maintenance work, some manufacturers have service contracts. Some progressive cities, states, and provinces require operators to be examined and licensed (see last chapter).

The second requirement is a log program. Next comes scheduling of maintenance, with one schedule for operation and a second one for shutdown. Both are vital. Also, set up shift schedules.

A log of temperatures, pressures, etc., provides an excellent diagnostic record to help spot troubles before they happen, just as temperature and blood pressure readings help a doctor. Machinery insurance companies pay out so many millions of dollars in claims yearly for wrecked machinery (due mostly to ignorance on the part of owners and operators) that some have set up preventive maintenance programs and log sheets for their clients. You can also obtain these programs from your equipment manufacturer. He wants his machinery to give satisfaction.

Q What information should be logged weekly if the compressor is a small reciprocating type (under 25 hp)?

A Check (1) the moisture indicator to determine any change in indicator chemical color or the presence of gas bubbles in the liquid refrigerant; (2) the oil sight glass to see that the level in the compressor is okay; (3) noise and vibration; and (4) temperatures. Make an inspection each week and place a check mark by each of the above four items *if* normal. In addition, jot down unusual operating conditions as they occur, and also the date when the evaporator and condenser filters are changed and other information. Date and initial each item.

MFGR.________ NO.________ SIZE ________ LOCATION ________ REFRIGERANT "R"________

Date	Amb. Temp	COMPRESSOR						MOTOR						CONDENSER				EVAPORATOR			
		Refrigerant				Oil		Volts			Amps.			Water		Air		Chiller		Fan Coil	
		Press.		Temp.				Phase			Phase			Temperature				Water Temp.		Air Temp.	
		Suct.	Disch.	Suct.	Disch.	Press.	Level	1	2	3	1	2	3	In	Out	In	Out	In	Out	In	Out

Fig. 21-1. Log sheet for reciprocating compressor of 25 hp and over.

Q Figures 21-1, 21-2, and 21-3 show log sheet data that should be recorded on some larger systems. In addition to these data, what other items must the daily log sheet indicate?

A Record the date and amount of oil or refrigerant added, repairs, adjustments, leak tests, service calls, shutdown on safety controls, water drained from purge (include amounts), and service calls.

PLANT ________ MACHINE SERIAL NO. ________ MACHINE SIZE ________

Time	COOLER					CONDENSER				COMPRESSOR						PURGE			Pump On or Off	Operator's Initials
	GPM ____					GPM ____				Bearing and Transmission Temperature	OIL				Motor Amps or Vane Position	Refrigerant Level	Condenser Pressure	Frequency of Pump Operation		
	Vacuum	Refrigerant Temperature	Refrigerant Level	Water Temperature		Vacuum or Pressure	Condensing Temperature	Water Temperature			Level	Temperature	Pressure	Reservoir Pressure						
				IN	OUT			IN	OUT											
1	2	3	4	5	6	7	8	9	10	11	12	13	14	15	16	17	18	19	20	21

Fig. 21-2. Log sheet for hermetic centrifugal machine.

Q When should the service man be called to avoid failure?

A Service is required when: (1) normal operating conditions are exceeded, (2) above-normal noise and vibration develop, (3) filters need cleaning or replacing, (4) an oil leakage occurs or the oil level changes, (5) the moisture indicator changes color, and (6) the bulls-eye shows a changed liquid flow pattern. These recordings become the system's history and enable you to set up a preventive maintenance program.

Q Work and shift schedules can be a problem if everyone isn't satisfied, especially in process industries which work around the clock. Set up a 168-hr-week work schedule for four operators.

MFGR.________ NO.________ SIZE ________ LOCATION ________ REFRIGERANT "R"________

DATE or Time	Air Temp.		COMPRESSOR							MOTOR		GEAR		CHILLER					CONDENSER				PURGE		INITIAL
			Position Cap. Indicator	Bearing Temp.		Oil						Oil		Refrigerant			Water Temp.		Refrigerant		Water Temp.				
	Ambient	Space				Level	Temp. Reservoir	Temp. Leaving Cooler	Pressure	Amps.	Volts	Pressure	Temp.	Level	Pressure	Temp.	In	Out	Pressure	Temp.	In	Out	Frequency of Operation	Water Removed	Watch or Shift Engineer

Fig. 21-3. Log sheet for centrifugal compressor.

A Fig. 21-4 is worked out to give a reasonable division of time and allow a post to be covered by four men, each working a 42-hr week. This plan works on a 32-hr cycle instead of the usual 24-hr (or daily) cycle, which always fights with the 7-day weekly cycle and thus causes disruption of habits in weekend change-over often requiring relief men.

Shift*	Day of week																				
	S	M	T	WT	F	S	S	M	T	WT	F	S	S	M	T	WT	F	S
First.......	A	B	C	D	A	B	C	D	A	B	C	D	A	B	C	D	A	B	C	D	A
Second....	D	A	B	C	D	A	B	C	D	A	B	C	D	A	B	C	D	A	B	C	D
Third......	C	D	A	B	C	D	A	B	C	D	A	B	C	D	A	B	C	D	A	B	C
First.......	B	C	D	A	B	C	D	A	B	C	D	A	B	C	D	A	B	C	D	A	B
Second....	A	B	C	D	A	B	C	D	A	B	C	D	A	B	C	D	A	B	C	D	A
Third......	D	A	B	C	D	A	B	C	D	A	B	C	D	A	B	C	D	A	B	C	D

*First, midnight to 8 AM; second, 8 AM to 4 PM; third, 4 PM to midnight.

Fig. 21-4. Four-man shift schedule based on 32-hr cycle.

Here are some of the advantages: (1) every man always has at least five full days (daylight) off every week; (2) no relief men are required; (3) every man always has three free Sunday day shifts out of four; (4) there are no long shifts *on*, or short periods *off*, and there is no weekly change of shifts with accompanying disruption of sleeping habits; (5) every man may have at least half the daytime off every day, and desirable sleeping time at least every other night; (6) no printed schedule is necessary, and bookkeeping is minimized; (7) the same men are always working together every day; (8) supervisors see every man at least once every four days; and (9) no man ever works over eight hours in any calendar day or in any consecutive 32 hours. While this schedule may be difficult to understand because it is so different from the conventional daily and weekly cycle, in one plant it was unanimously preferred by the men after they tried it.

NOTE: For 12 other work schedules designed to meet almost every situation, see chapter 18, *Standard Boiler Operators' Questions and Answers*, in your library.

SUGGESTED READING

Publications (magazines)
ASHRAE Journal
Airconditioning, Heating & Refrigeration News
Heating, Piping, Air Conditioning
Power

22

LICENSE REQUIREMENTS
FOR
REFRIGERATION AND
AIR CONDITIONING ENGINEERS
IN THE
UNITED STATES AND CANADA

U.S. REFRIGERATION OPERATING ENGINEER'S LICENSE REQUIREMENTS

States and cities	Examination required	Education, experience, and remarks	Class license	Citizenship	Local residence	Nonlocal licenses recognized	Examination fee and renewal	Minimum refrigeration capacity requiring licensed operator
Alabama	No state license							
Alaska	No state license							
Arizona	No state license							
Arkansas	No state license							
California	No state license							
Colorado	No state license							
Denver: Board of Examiners, City and County Bldg., Denver 80202	Yes, written; must know refrigeration code, principles, all type refrigerants	Age (not specified), (a) apprentice training, or (b) recognized trade school, or (c) experience, or (d) mechanical engineer's degree plus 1 year experience	(1) Stationary engineer Class A (2) SR-1	Not required	Not required	No	$10.00, $5.00 annually	(1) Refrigeration system of Group 2 or Group 3 refrigerants, having charges of 200 lb or more; or (2) System having manual or semiautomatic controls, with charges of 400 lb or more of Group 1 refrigerants; or (3) System with fully automatic controls, with charges of 1,500 lb or more of Group 1 refrigerants Note: semiautomatic shall mean plants or systems with automatic safety controls but manual load reportioning controls requiring other than seasonal adjustments Refrigerant equipment of

Connecticut New Haven: Board of Examiners of Engineers, City Hall, Church Street, New Haven	No state license	Yes, oral	Age (not specified), prove qualification for machinery he is to operate	(1) Refrigeration operator	Not specified	Required	Yes	$5.00, $2.00 annually	All refrigeration or air-conditioning systems having 100 lb of refrigerant or over	fully automatic controls, with charges between 750 and 1,500 lb of Group 1 refrigerants
Delaware	No state license								
District of Columbia Dept of Occupations and Professions 1145 19th St., N.W. Washington, D.C.		Yes, written; sketches and practical	Age 21, at least 1 year experience in care and operation of refrigeration equipment	(1) Refrigeration operating engineer, Class 8-A	Not specified	Not specified	No	$3.00, $3.00 bi-annually	To take charge of and operate refrigeration systems where the total connected load is not in excess of 200 compressor horsepower or 200 tons for an absorption system and where there is no other equipment requiring a licensed steam or other operating engineer. (a) Compressor horsepower is the manufacturer's horsepower rating given on nameplate, (b) ton of refrigeration is the removal of 12,000 Btu per hr, (c) horsepower rating of a refrigeration system shall be either compressor horsepower or tons of refrigeration as converted to horsepower on an equal basis

U.S. REFRIGERATION OPERATING ENGINEER'S LICENSE REQUIREMENTS (Continued)

States and cities	Examination required	Education, experience, and remarks	Class license	Citizenship	Local residence	Nonlocal licenses recognized	Examination fee and renewal	Minimum refrigeration capacity requiring licensed operator
Florida	No state license							
Tampa: Department of Public Works, Plumbing, Gas, and Boiler Bureau, City Hall Tampa 2	Yes, written; also examined in electricity and diesel engines	Age 21, (a) employed as oiler or assistant under first-class engineer for not less than 3 years, or (b) served as fireman or assistant to engineer on steamboat or railway locomotive for not less than 3 years and employed for not less than 1 year as assistant under licensed first-class engineer, or (c) journeyman, machinist, or boilermaker with 2 years' experience at time served in apprenticeship and over 2 years' experience as assistant under first-class engineer, or (d) graduate of mechanical engineering with over 1 year experience under first-class engineer, or (e) holder of second-class engineer's license issued by city of Tampa with 2 years' experience on license, or (f) holder of first-grade or first-class engineer's license of any U.S. city having over 100,000 population or holding steam engineer's license of U.S. government or marine service	(1) First-class refrigeration engineer, in charge of all refrigeration machinery driven by diesel, internal combustion, or electricity	Not required	Not required	Yes, but as proof of experience, examination must be taken	$25.00, $3.00 annually	All refrigeration plants of over 50-ton capacity in which refrigerant used comes within refrigerant defined as Group 2 and Group 3, as revealed in Section 5, ASA B9, 1939, ASHRAE Standard Safety Code for Mechanical Refrigeration, and subsequent revisions

Georgia	No state license							
Hawaii	No state license							
Idaho	No state license							
Illinois	No state license							
Indiana	No state license							
Iowa	No state license							
Kansas	No state license							
Kentucky	No state license							
Louisiana	No state license							
New Orleans: Department of Safety and Permits, Board of Examiners of Operating Engineers,	Yes, written and oral	Age 21, at least 2 years' experience in refrigeration plants	(1) First-class refrigeration engineer	U.S. citizen	At least 6 months	Yes, but as proof of experience, examination must be taken	$6.50	Plants with over 150-ton capacity
	Yes, written; also examined in electricity and diesel engines	Not specified	(2) Second-class refrigeration engineer, in charge of all refrigeration machinery driven by diesel, internal combustion, or electricity up to 175 hp	Not required	Not required	Yes, but as proof of experience, examination must be taken	$15,00, $3,00 annually	Plants from 30- to 50-ton capacity in which refrigerant used comes within the refrigerant classification defined as Group 2 and Group 3 as revealed in Section 5, ASA B9, 1939, ASHRAE Standard Safety Code for Mechanical Refrigeration, and subsequent revisions

U.S. REFRIGERATION OPERATING ENGINEER'S LICENSE REQUIREMENTS (Continued)

States and cities	Examination required	Education, experience, and remarks	Class license	Citizenship	Local residence	Nonlocal licenses recognized	Examination fee and renewal	Minimum refrigeration capacity requiring licensed operator
Louisiana (cont'd) Mechanical Inspection Section, City Hall, New Orleans								
	Yes, written and oral	Age 18	(2) Second-class refrigeration engineer	U.S. citizen	At least 6 months	Yes, same as above	$5.50	Plants with up to 150 tons, or plants of unlimited tonnage if plant is in charge of a first-class engineer
	Yes, written and oral	Age 18	(3) Third-class refrigeration engineer	U.S. citizen	At least 6 months	Yes, same as above	$4.00	Plants with up to 75-ton capacity, and limited to plants of not over 75-ton capacity
	Yes, written and oral	Age 18	(4) Special class refrigeration engineer	U.S. citizen	At least 6 months	Yes, same as above	$3.00	Plants with up to 20-ton capacity using Group 1 refrigerants; plants of 5 to 20 tons when using Group 2 refrigerants
Maine	No state license							
Maryland	No state license							
Massachusetts	No state license							
Michigan Detroit: Dept. of Bldg's, and Safety Engineering,	Yes, written and oral	Age 21, (a) at least 2 years' experience operating refrigerating equipment, or (b) degree in mechanical or refrigeration engineering	(1) First-class refrigeration operator	Not required	At least 30 days	No	$8.00, $4.00 annually	Plants of over 250 hp, or having over 1,000 lb of refrigerant

State / Location	Examination	Requirements	Class of license	Citizenship	Residence		Fee	Plants covered
Fourth Floor, City-County Bldg., Detroit 48226	Yes, written and oral	accepted in lieu of 1 of the required 2 years' experience / Age 21, (a) 2 years' experience operating refrigerating equipment, or (b) degree in mechanical or refrigeration engineering accepted in lieu of one of the required 2 years' experience	(2) Second-class refrigeration operator	Not required	At least 30 days	No	$8.00, $4.00 annually	Plants of over 125 to 250 hp, or having over 500 to 1,000 lb of refrigerant
	Yes, written and oral	Age 21, minimum experience but must be familiar with refrigeration equipment	(3) Third-class refrigeration operator	Not required	At least 30 days	No	$4.00, $3.00 annually	Plants of over 25 to 125 hp, or having 100 to 500 lb of refrigerant
Minnesota	No state license							
Missouri								
Kansas City: Division of Building and Inspection, City Hall, Kansas City 64106	Yes, written and oral	Age 21, at least 5 years' experience in steam and refrigeration plants. Degree in mechanical engineering counts for 2 years' experience	(1) Class A (first grade) operating engineer	Not specified	Not specified	Not specified	$7.50, $5.00 annually	Plants with capacity of over 10 tons of Group 2 or Group 3 refrigerant
	Yes, oral and written	Age 21, at least 3 years' experience in operation. Degree in mechanical engineering counts as 1 year experience	(2) Class B refrigeration operating engineer	Not specified	Not specified	Not specified	$7.50, $5.00 annually	Plants with capacity of over 10 tons of Group 2 or Group 3 refrigerant
St. Joseph: Board of Engineers, City Hall, St. Joseph	Yes, oral or written	Age 21, at least high school	Refrigeration operator	U.S. citizen	Required	No	$7.50, $2.00 annually	All plants of 25 tons or over
Montana	No state license							

U.S. REFRIGERATION OPERATING ENGINEER'S LICENSE REQUIREMENTS (Continued)

States and cities	Examination required	Education, experience, and remarks	Class license	Citizenship	Local residence	Nonlocal licenses recognized	Examination fee and renewal	Minimum refrigeration capacity requiring licensed operator
Nebraska	No state license							
New Jersey Mechanical Inspection Bureau, Department of Labor and Industry, John Fitch Plaza, Trenton 08625	Yes, state license Yes, oral, written, or both	Age 18, (a) must have Class B license plus 1 year's service in charge of Class B plant, or (b) 2 years' experience operating a plant under Class A engineer	(1) Class A refrigeration (unlimited capacity)	U.S. citizen	Conditional	Yes, but for endorsement of application only	$10.00, $3.00 annually	(1) May act as chief engineer in any plant
	Yes, oral, written, or both	Age 18, must have Class C license and 1 year practical experience as engineer on license, or equivalent	(2) Class B (second grade) refrigeration	U.S. citizen	Conditional	Yes, but for endorsement of application only	$10.00, $3.00 annually	(1) May act as chief engineer of plants having not over 300 tons refrigerating capacity; or (2) May act as operator in larger plants, but under supervision of properly licensed chief engineer
	Yes, oral, written, or both	Age 18, if refrigerant in plant is inflammable or toxic, must have had 1 year experience as oiler or assistant, or 6 months as an operator of such equipment	(3) Class C (third grade) refrigeration	U.S. citizen	Conditional	Yes, but for endorsement of application only	$10.00, $3.00 annually	(1) May act as chief engineer of plants having not over 65 tons refrigerating capacity; or (2) May act as operator in larger plants, but under supervision of properly licensed chief engineer; or (3) Any refrigerating system using inflammable or toxic refrigerant and rated over 6 tons must be in charge of licensed operator
New Mexico	No state license							

State/Agency	State license	Examination	Experience requirements	Grade	Citizenship		Reciprocity	Fees	Scope
New York Buffalo: Dept. of Public Works, Examiner of Stationary Engineers and Inspection of Steam Boilers, City Hall, Buffalo	No state license	Yes, written and practical. Any person holding an engineer's license under the provisions of the city of Buffalo Stationary Engineers Ordinance may also hold any grade refrigeration operator's license, for which he can qualify by written and practical examination of his qualifications and experience of refrigeration power developing units	Age 21, 3 years' practical experience as first-class refrigeration operator or as a first-class stationary engineer	(1) Chief refrigeration operator	Declared intention	Not indicated	Yes, if equal in requirement to Buffalo license	$25.00, $10.00 annually	Refrigeration power developing units of any horsepower
		Yes, written and practical	Age 21, 2 years' practical experience as a second-class refrigeration operator or a second-class stationary engineer	(2) First-class refrigeration operator	Declared intention	Not indicated	Yes, if equal in requirement to Buffalo license	$25.00, $7.50 annually	Up to 225 hp of refrigeration power developing units
		Yes, written and practical	Age 21, (a) 3 years' practical experience in operations or repairs of refrigeration power developing units, or (b) graduate from recognized technical school of engineering and 1 year practical experience on refrigeration power developing units	(3) Second-class refrigeration operator	Declared intention	Not indicated	Yes, if equal in requirement to Buffalo license	$25.00, $5.00 annually	Up to 150 hp of refrigeration power developing units *Exceptions and conditions:* Unitary or self-contained air-conditioning units used for domestic, commercial, or industrial purposes, singly or combined, shall be exempt from the provisions of this ordinance. All units over 1/2 hp singly or combined; up to 60-hp capacity, using ammonia or carbon dioxide refrigerants; or up to 100-

States and cities	Examination required	Education, experience, and remarks	Class license	Citizenship	Local residence	Nonlocal licenses recognized	Examination fee and renewal	Minimum refrigeration capacity requiring licensed operator
New York (cont'd)								hp capacity, using Freon or similar refrigerants, shall be equipped with safety devices. Licensed operator is needed for device, apparatus, or machine used singly or combined to operate ammonia or carbon dioxide refrigerating units over 60 hp or over 100 hp using Freon or similar refrigerant
Mount Vernon: Boiler and Refrigeration Operator and Inspector, 470 East Lincoln Ave., Mt. Vernon	Yes, oral and written	Age 21, no stated requirements	Refrigeration operator	Not required	Not required	Not specified	$5.00, $5.00 annually	All refrigeration units containing 200 lb and over of refrigerant must have licensed operator
New York City: Department of Personnel, Civil Service Commission, Municipal Bldg., 229 Broadway, New York 10007	Yes, written and practical	Age 21, no specific education or experience required, must read and write English	Refrigeration machine operator, unlimited capacity	U.S. citizen	At least 3 years	No	$10.00	(1) Refrigeration system containing Group 1 refrigerant (other than CO_2) if (a) installed before June 1, 1957, and contains over 50 lb of refrigerant or (b) installed after June 1, 1957, and has prime mover or compressor of over 50 hp (2) Any combination of re-

frigerating system for (a) human comfort air conditioning, (b) under sole direct control of single occupant, (c) containing a Group 1 refrigerant (other than CO_2) and prime movers and compressors total more than 100 hp. All industrial plants, other than for human comfort, containing more than 50 lb of a Group 1 refrigerant

Location	Examination	Requirements	License	Citizenship			Fee	Scope
Rochester: Public Safety Bldg., Room 200, Civic Center Plaza, Rochester 14614	Yes, oral, written, or both	Age 21, Grade R-2 license and 5 years' experience	Grade R-1 refrigeration operator	U.S. citizen	Not specified	Not specified	$15.00, $10.00 annually	Operate any refrigeration plant, any tonnage, other than steam motive power
	Yes, oral, written, or both	Age 21, Grade R-3 license and 3 years' experience	Grade R-2 refrigeration operator	U.S. citizen	Not specified	Not specified	$12.00, $7.00 annually	Operate any refrigeration plant not over 1,000 tons, other than steam motive power
	Yes, oral, written, or both	Age 21, Grade R-4 license and 2 years' experience	Grade R-3 refrigeration operator	U.S. citizen	Not specified	Not specified	$10.00, $5.00 annually	Operate any plant not over 750 tons, other than steam motive power
	Yes, oral, written, or both	Age 21, 1 year experience	Grade R-4 refrigeration operator	U.S. citizen	Not specified	Not specified	$8.00, $3.00 annually	Operate any plant not over 400 tons, other than steam motive power
Tonawanda: City Hall	Yes, written and practical	Age 21, 3 years' experience as first-class refrigeration operator or first-class stationary engineer	Chief refrigeration operator	U.S. citizen	Yes	Yes	$10.00, $10.00	Operate any hp units, other than steam-driven
	Yes, written and practical	Age 21, 2 years' experience as second-class refrigeration operator or second-class stationary engineer	First-class refrigeration operator	U.S. citizen, or intention filed	Yes	Yes	$7.50, $7.50 annually	Operate 250-hp units, other than steam-driven

U.S. REFRIGERATION OPERATING ENGINEER'S LICENSE REQUIREMENTS (Continued)

States and cities	Examination required	Education, experience, and remarks	Class license	Citizenship	Local residence	Nonlocal licenses recognized	Examination fee and renewal	Minimum refrigeration capacity requiring licensed operator
New York (contd)	Yes, written and practical	Age 19, 3 years' experience repairing or operating similar equipment, or 1 year experience if graduate of technical school	Second-class refrigeration operator		Yes	Yes	$5.00, $5.00 annually	Operate 150-hp units, other than steam-driven
White Plains: Fire Department, 219 Mamaroneck Ave., White Plains 10605	Yes, written	Age 18, if found competent by examination	Refrigeration operator	Not specified	Not required	Not required	$5.00	Operate refrigeration system of over 50-ton capacity
North Carolina	No state license							
North Dakota	No state license							
Ohio	No state license							
Oklahoma	No state license							
Oregon	No state license							
Pennsylvania	No state license							
Erie: Department of Public Safety, Bureau of Air Pollution and Engineers	Yes, written	Age 21, 2 years' experience in refrigeration plants or technical school training	Class 6 license, refrigeration operator	U.S. citizen	Not required	Not specified	$5.00, $1.00 annually	Licensed operator required in all refrigeration plants of 50 hp or over

	Examination	Experience	License	Citizenship	Bond	Reciprocity	Fee	Remarks
Licenses, City Hall Annex, Erie								
Philadelphia: Bureau of Building Inspection, 224 City Hall Annex, Philadelphia 19107	Yes, written	Age 21, 2 years assisting in refrigerating plants	Refrigeration engineer, Class B	Not required	Not required	No	$4.00, $3.00 annually	Licensed operator required in all refrigeration plants of 25 tons or if using refrigerant which is toxic, odorous, tastable, flammable, or explosive
Pittsburgh: Bureau of Building Inspection, 100 Grant St., Pittsburgh 15219	Yes, written	Age 21, 2 years' experience in refrigeration plants	(1) Class A refrigeration operator	U.S. citizen	Not required	No	$15.00, $5.00 annually	Licensed operator required in all commercial refrigeration plants or those in public buildings regardless of tonnage
	Yes, written	Age 21, 2 years' experience in refrigeration plants	(2) Class B refrigeration operator	U.S. citizen	Not required	No	$15.00, $5.00 annually	
Rhode Island	No state license							
Providence: Department of Public Service, City Engineer, City Hall, Providence 02903	Yes, oral	Age 21, at least 6 months' work under supervision of licensed refrigeration operator	Refrigeration machine operator	U.S. citizen	Not required	No	$5.00, $2.00 annually	(1) Licensed operator required in all plants of 100 tons, using Group 1 refrigeration; (2) Plants of 15 tons using Group 2 refrigeration
South Carolina	No state license							
South Dakota	No state license							
Tennessee	No state license							

307

U.S. REFRIGERATION OPERATING ENGINEER'S LICENSE REQUIREMENTS (Continued)

States and cities	Examination required	Education, experience, and remarks	Class license	Citizenship	Local residence	Nonlocal licenses recognized	Examination fee and renewal	Minimum refrigeration capacity requiring licensed operator
Tennessee (cont'd)								
Memphis: Board of Examiners, Stationary Engineers, City Hall, Memphis	Yes, written	Age 21, at least 3 years' experience in refrigeration plants	(1) First-class refrigeration operator	Not required	Required	No	$20.00, $10.00 annually	Licensed operator required in all refrigeration plants of 40-ton capacity
	Yes, written	Age 21, at least 3 years' experience in refrigeration plants	(2) Second-class refrigeration operator	Not required	Required	No	$20.00, $8.00 annually	
	Yes, written	Age 21, at least 3 years' experience in refrigeration plants	(3) Third-class refrigeration operator	Not required	Required	No	$20.00, $4.00 annually	
Texas	No state license							
Utah	No state license							
Vermont	No state license							
Virginia	No state license							
Washington	No state license							
Seattle: Civil Service Department, 500 Union Pacific Bldg., 1000 Second Ave., Seattle 98104	Yes, oral and written	Age 18, no specific time period required. Must know type of equipment where applicant expects to work or be familiar with refrigeration plants	Refrigeration operating engineer	Not required	Not required	Not specified	$7.50 for examination, $5.00 license fee, $5.00 annually	No tonnage requirements' but anyone who performs repairs where it is necessary to open refrigerant-containing parts must be licensed

West Virginia	No state license
Wisconsin	No state license

CANADIAN REFRIGERATION OPERATING ENGINEER'S LICENSE REQUIREMENTS

Province	Examination required	Education, experience, and remarks	Class license	Citizenship	Local residence	Nonlocal licenses recognized	Examination fee and renewal	Minimum refrigeration capacity requiring licensed operator
Alberta	No license							
British Columbia	No license							
Manitoba: Mechanical & Engineering Division, Room 611, Norquay Bldg., Winnipeg 1	Yes, written	Age 18, 1 year experience in 50-hp plant, or in related trade and assisted in 50-hp refrigeration plant for 6 months	Refrigeration Plant Operator	Not required	Not required	Up to Board of Examiners	$7.00, $2.00 annually	Charge of plant with hp rating not over 500, or act as shift operator in any plant. All refrigeration plants over 50 hp of over 15 psi, except factory-assembled units not over 100 hp of nontoxic nonflammable refrigerant, require license
New Brunswick	No license							
Newfoundland and Labrador Boiler Inspection Branch, Confederation Bldg., St. John's	Yes, written	Age 19, 1 year experience in plant of over 25 hp on shift in compressor room	Grade B refrigeration operator	Not specified	Not specified	Not specified	$5.00, $2.00 annually	Act as chief operator of stationary refrigeration plant not exceeding 400 hp, or as shift operator in plant of unlimited hp
	Yes, written	Age 21, 1 year qualification for Grade B license and 2 years in stationary refrigeration plant of over 300 hp on shift in compressor room	Grade A refrigeration operator	Not specified	Not specified	Not specified	$7.00, $2.00 annually	Act as chief, or shift operator in charge, of refrigeration plant of unlimited hp

CANADIAN REFRIGERATION OPERATING ENGINEER'S LICENSE REQUIREMENTS (Continued)

Province	Examination required	Education, experience, and remarks	Class license	Citizenship	Local residence	Nonlocal licenses recognized	Examination fee and renewal	Minimum refrigeration capacity requiring licensed operator
Northwest Territories Boilers and Pressure Vessels Inspection, Yellowknife, N.W.T.	Yes, but only if cannot prove qualification	Qualified to take charge of any plant requiring Class I license	Class I operating engineer (All N.W.T. licenses are combination steam and refrigeration)	Not specified	Not specified	Not specified	Not specified	Act as chief engineer of any plant with boilers or pressure vessels (containing refrigerant)
	Yes, but only if cannot prove qualification	Qualified to take charge of any plant requiring Class II license	Class II operating engineer	Not specified	Not specified	Not specified	Not specified	Take charge of boiler, pressure vessel, or plant not over 750 hp, or operate similar equipment of any capacity as shift engineer under chief engineer
	Yes, but only if cannot prove qualification	Qualified to take charge of any plant requiring Class III license	Class III operating engineer	Not specified	Not specified	Not specified	Not specified	Take charge of boiler, pressure vessel, or plant not over 500 hp, or operate similar equipment of 750 hp as shift engineer under chief engineer
	Yes, but only if cannot prove qualification	Qualified to take charge of any plant requiring Class IV license	Class IV operating engineer	Not specified	Not specified	Not specified	Not specified	Take charge of boiler or pressure vessel not over 200 hp, or operate similar equipment of not over 500 hp under Class III engineer
	Yes, but only if cannot prove qualification	Qualified to take charge of any plant requiring Class V license	Class V operating engineer	Not specified	Not specified	Not specified	Not specified	Take charge of low-pressure boiler or pressure vessel not over 50 hp, or operate similar equipment of 500 hp under Class III engineer

Nova Scotia: Department of Labor, Johnston Bldg., Halifax	Yes, written	Age 18, 24 months' experience, of which 12 months with equipment of over 200 mhp	First-class refrigeration operator	Canadian citizen	Not required	After 1 year in province	$5.00, $1.00 annually	Any refrigeration plant having driving units of 50 hp or over
	Yes, written	Age 18, 18 months' experience, of which 9 months with equipment of over 100 mhp	Second-class refrigeration operator	Canadian citizen	Not required	After 1 year in province	$4.00, $1.00 annually	Not stipulated
	Yes, written	Age 18, 9 months' experience with equipment of over 15 mhp	Third-class refrigeration operator	Canadian citizen	Not required	After 1 year in province	$4.00, $1.00 annually	Not stipulated
Ontario: Board of Examiners of Operating Engineers Branch, 400 University Ave., Toronto 2	Yes, written; score not less than 60 per cent on examination	Age 23, must be able to read, write, and sketch, (a) hold Class B refrigeration operators certificate, and (b) have had at least 4 years in a refrigeration plant, of which not less than 3 years served in plant of over 400 hp	(1) Refrigeration operator, Class A	Not required, one year's residence and declaration of intention of becoming Canadian citizen	Not required	Not stipulated	$10.00, $8.00 annually	Act as chief engineer in refrigeration plant of unlimited tonnage. Note: All plants of 51 registered horsepower or over must be in charge of licensed engineer
	Yes, written; score not less than 60 per cent on examination	Age 19, (a) at least 1 year experience in refrigeration plant, or (b) 18 months' experience installing and servicing equipment in refrigeration plants	(2) Refrigeration operator for Class B	Same as above	Not required	Not stipulated	$8.00, $5.00 annually	Act as chief engineer in any plant having refrigeration capacity between 50 and 400 registered horsepower
Prince Edward Island	No license							
Quebec: Board of Examiners, 355 McGill St., Montreal 1	Yes, written, score at least 70%	Not stipulated	Class A stationary engineman, refrigeration	Not stipulated	Not stipulated	Not stipulated	$10.00, $5.00 annually	Operate or take charge of any refrigeration installation

Province	Examination required	Education, experience, and remarks	Class license	Citizenship	Local residence	Nonlocal licenses recognized	Examination fee and renewal	Minimum refrigeration capacity requiring licensed operator
Quebec (contd)								
	Yes, written, score at least 70%	Not stipulated	Class B stationary engineman, refrigeration	Not stipulated	Not stipulated	Not stipulated	$8.00, $4.00 annually	Operate refrigeration apparatus of not over 300 mhp of Groups II and III refrigerants, or of 1,200 mph of Group I, or take charge of any Class A installation
	Yes, written, score at least 60%	Not stipulated	Class C stationary engineman, refrigeration	Not stipulated	Not stipulated	Not stipulated	$5.00, $2.00 annually	Operate refrigeration apparatus of not over 200 mhp of Groups II and III refrigerants, or of 800 mhp of Group I, or take charge of any Class B installation
	Yes, written, score at least 60%	Not stipulated	Class D stationary engineman, refrigeration	Not stipulated	Not stipulated	Not stipulated	$3.00, $2.00 annually	Operate refrigeration apparatus of not over 100 mhp of Groups II and III refrigerants, or of 400 mhp of Group I, or take charge of any Class C installation
Saskatchewan Department of Labor, Boiler Pressure Vessel and Elevator Branch, Regina	Yes, written	Age 18, (a) At least 12 months' experience operating or assisting in operation of a refrigeration plant of not less than 3 tons capacity per 24 hr, or (b) is holder of first-, second-, third-, or fourth-class engineer's certificate and has for a period of 6 months operated or assisted in operation of refrigeration of not less than 3 tons refrigeration per 24 hr	Refrigeration engineer's certificate	Not required	Not required	Not stipulated	$5.00, $1.00 annually	All plants having a capacity of more than 10 tons per 24 hr

AUTHOR'S NOTE: If your state, province, county, or city has license requirements and it is not listed on this chart, kindly write us. We shall include your standards in the next revision of this book as these are the only data on refrigeration license requirements published anywhere. For stationary engineer license requirements, see Steve Elonka and Joseph F. Robinson, Standard Plant Operator's Questions and Answers, Vol. II, McGraw-Hill Book Company, New York.

INDEX

INDEX

Absolute pressure, 7
Absorption systems, 17–21, 234–236
  advantages of, 21
  ammonia-water, 21
  lithium bromide, 21
  steam supply, 182
Accumulator, 58
Air:
  blending, 244
  definition of, 228
  liquification, 288
  mixtures, 241
  moist, enthalpy of, 249
  in system, 183
  temperature versus humidity, 228
Air conditioning, 227–262
  central system, 244
    advantages of, 245
  complete, 227, 258
  compression cycle, 237
  computerized control, 259
  dust removal, 240
  electric lights, effect of, 193
  factors effecting load, 227
  high-velocity system, 145
  ice system, 243
  infiltration effect, 228
  packaged unit, 241
  people load, effect of, 227
  planning of, 241
  refrigerant expansion, 243
  shutdown for season, 256
  starting centrifugal unit, 251
  sunlight, effect of, 227
  temperature control, 238
  troubleshooting, calculations
    for, 247–251
  use of ground water in, 243
  zoning, 245

Algae, 225, 265, 266
Ammonia, 15, 21, 45
  burns, 276
  charging, 178, 179, 180
  coils, bottom-fed, 70
  dangers of, 277
  dump connection, 279, 280
  exposure limits, 276
  lube oil characteristics of, 110
  safety precautions for, 275
  size of storage drums, 178
  slugs in system, 280
  withdrawing of, 179, 180
Approach, 209
Automatic expansion valve, 73, 74
  precautions for, 75

Balancing load, 186
Baudelot cooler, 70
Bearings, ring-oiled, 133
Blowdown, 210
Boilers, 294
Boiling point, 1, 28
Booster cycle, 23, 24
Brine, 268
  acidity test, 270
  ammonia leakage into, 167, 269
  calcium chloride, 269
  calculations, data for, 166
  coolers, 268
  density, 268
  qualities needed, 269
  sodium chloride, 269
  testing, 270
British thermal unit (Btu), 1
Bypass capacity control, 46

Calculations, 159–175
  adding calcium chloride to brine, 166
  for air conditioning, 247–251
  Btu needed, 160
  capacity of compressor, 171
  coefficient of performance, 159
  condenser cooling water, 164
  cooling-tower blowdown, 267
  cooling water per ton, 165
  displacement of compressor, 168
  heat: generated in cold rooms, 164
    leaking into cold room, 161–164
    removing from cold spaces, 165
  ice-making versus refrigeration tonnage, 166
  ice-making capacity, 154
  K factor, 162
  lost cooling effect, 171
  refrigerating effect of R-12, 165
  time needed for cooling, 161
Capillary tube, 73
Carbon dioxide (CO_2), 28, 45, 278
  dangers of, 277
Carbon tetrachloride, dangers of, 277
Cascade system, 25, 289
Centrifugal compressor systems, 251–256
  automatic damper control, 254
  condenser water control, 253
  shutdown periods, 256
  starting of, 251
    hand operating controls, 253
    purge-recovery, 252
    steam turbine drive, 252
  surging operation, 252
Centrifugal system, 25
Charge connection, 117
Charging ammonia, 178
  Freon-12, 180
  refrigerant needed, 179
Chemicals, 265, 266
Claude system, 289
Cleaning solvents, 277, 278
Clearance pocket, 46
Codes, 224, 242
  city, local, 243, 274
  drainage requirements, 242
  refrigeration, 272
Compounding, 24
Compression ratio (CR), 17
Compression systems, 15
  pressures in, 17

Compression systems (*Cont.*):
  temperatures for, 16
Compressor clearance, 45
Compressor motors and drives, 146–158
Compressor output, 171
  securing, 258
Compressors, 37–54
  angle, 41
  booster, 41
  calculations, data for, 169, 170
  capacity: control 46, 245
    reduction, 47
  centrifugal, 25, 37
    lubrication, 134
    starting, 251
    surging in, 182
  crankcase explosions, 273
  helical rotary-screw (hrs), 38, 39
    capacity control, 50, 52
    dry type, 50
    of slide valve, 51
    in separating oil, 52
  hermetic type, 42, 53
  horizontal double-acting (hda), 39
  multiple-effect, 42
  reciprocating, 37, 39
    starting, 181
  rotary sliding vane, 37
  V-type, 41
  vertical single-acting (vsa), 39, 40
    removing cylinder head, 189
  VW-type, 42
Condenser, 15, 16, 55–65
  air-cooled, 55, 56
  atmospheric, 55, 60
  cleaning, 64
  double-pipe, 55, 59
  evaporative, 55, 60–62, 216
    capacity of, 61, 63, 250
    maintenance of, 62
    operation of, 62
    securing, 258
  'selection of, 55
  shell-and-coil, 55, 58
  shell-and-tube, 55, 56, 257
    horizontal closed, 58
    vertical open, 56, 58
  water-cooled, 56
    gallon-degrees per minute per ton, 56
    water per ton capacity, 56

Controls:
  automatic, 101
  differential, 102
  electric, 96, 106
    precautions for, 102
  flow type, 96, 101
  pressure: bellows, 96
    Bourdon tube, 96, 97
    diaphragm type, 96, 97
  reset pressure: automatic, 101
    manual, 101
  safety, 103
  temperature, 97, 98
    differential, 98
    range, 98
  testing, 92
Cooler, hookup, 111
Cooling effect, 15
Cooling range, 209
Cooling towers, 207–226
  algae growth, 225
  approach, 209
  blowdown, 267
  classification of, 211
  delignification of, 266
  dissipating heat, 207
  drift, 210
  electric motors for, 225
  versus evaporative condenser, 221
  fogging, 220
  forced-draft type, 213
  hyperbolic, 217
    operation of, 218
  icing effect, 220
  increasing cooling effect, 208, 222
  induced-draft: counterflow type, 214
    double-flow, 214
    ice formation in, 219, 220
  installed indoors, 217
  laying up, 225
  limitations of, 221
  maintenance of, 221
  mechanical-draft type, 212
  pump, sizing of, 222
  quality air, 208
  reversing fan, 220
  shutting down, 221
  spotty performance of, 219
  spring startup, 224
  summer operation, 224
  system, 215
  winter operation, 220
    protection from freezing, 223

Cooling towers:
  wood destruction, 266
Cooling water, 263–271
  scale formation, 264
  systems, 263
    circulating, 264
    closed, 264
  treatment for, chemical, 263
Critical pressure, 8, 28
Critical temperature, 8, 28
Cryogenics, 287–291
  instruments for, 290
  liquification system, 291
Cutouts:
  high-pressure, 100, 107
  low-pressure, 100, 107
Cylinder head, removing, 189

Dalton's law, 7
Defrosting, 141–145
  ammonia coils, 142, 143
  with electric resistance heaters, 143
  with full automatic device, 144
  with hot discharge gas, 144
  with water, 143
Degree-days, 11
Dew point, 229
Direct-expansion system, 69
Discharge line, 116
Drafts, preventing, 246
Driers, 188
Drift, 210
Dry-bulb temperature, 208
Dry ice (CO_2), 205, 278

Electric drives, 157
Electric motors (*see* Motors, electric)
Ethylene, 288
Evaporator, 66–71
  Baudelot cooler, 70
  dry-type, 66
  extended-surface, 67
  flooded-type, 66
  heat-transfer ratio, 70
  plate coil, 68
  prime surface, 66, 67
  shell-and-tube, 68
  static head in, 70
Expansion engine, 288

Expansion valves, 15, 16, 72
  automatic, 73, 74, 77
    precautions for, 75
  capillary tubes, 73
  hand-type, 73
  high-side floats, 73, 81–83
  low-side floats, 73, 81, 82
  malfunction of, 185
  thermostatic, 73, 75–77, 80
    equalizers of, 77
    external equalizers on, 78, 79
    multi-outlet, distributor, 79
    remote bulb, 79
  (*See also* Valves)

Fahrenheit, 1, 2
Filling in cooling towers, 211
Fire extinguishers, 274
Fittings (*see* Piping and fittings)
Float switch, 83
Floats, high-side, 83
Food storage, 199
  high humidity in, 201
Foul gas, 63
Fouling factor, 64
Freezing, 14
  explosions caused by, 274
Freezing process:
  blast, 200
  contact, 200
  immersion, 200
  quick, 199
  sharp, 199
Frigorific mixture, 8
Fruit, precooling, 203
Fungus attack, 266, 267

Gage glass, safety feature of, 273
Gage pressure, 7
Gallon-degrees per minute per ton, 56
Gas masks, 274, 281, 282
Gaskets, 136

Halide torch, 187
Halocarbons, 35
Hand expansion valve, 73
Heat, 1
  absorbing capacity of matter, 9
  kinetic theory of, 6

Heat (*Cont.*):
  latent, 4, 28
  of respiration, 203
  sensible, 3
  specific, 2, 3
  vaporization of, 4, 28, 29
Heat load, 209
Heat pump, 237, 238
Heat transmission, 227
Hermetic compressor, 25, 26, 53
High-side floats, 73, 74
Horsepower equivalent, 159
Hydrogen, 29, 288

Ice, 4, 14
  artificial, 203
  in cooling towers, 219, 220
  dry, 205
  making, 204, 205
  in refrigeration, 14
  white versus clear, 204
Insulation, thermal, 117–123
  conductivity K-factor, 119
  materials, 121
  moisture in, 119
  properties of, 118
  reflective, 123
  vapor barrier, 121
Interlocked systems, 283

Jet cooler, 234
Joints, flanged, 108
Joule-Thomson effect, 288

Langelier index, 264
Lantern ring, 139
Latent heat, 4, 28
Leak in system, 156, 282
License requirements for operators in
    the United States and Canada,
    295–312
Lights, open-flame, 282
Liquid coolers, 71
  heat-transfer rates, 71
Log sheets, 292, 293
Low voltage as a killer, 278
Lubrication, 44, 124–140
  adding, 44, 132
    to bearings, 127

Lubrication (*Cont.*):
  care needed, 132, 134
  characteristics of, 128, 129
  checks needed, 284
  cloud point, 130
  consumption, excessive, 132
  for cylinders, 126, 132
  deposits, cause of, 131, 132
  draining from compressor, 133
  flash point, 130
  foaming in compressor, 127
  forced-feed system, 124, 126
  halocarbons, effect on, 130
  overexpansion in crankcase, 127
  oxidation, 127
  pour point, 130
  precautions for, 126
  problems, evaporator, 125
  sludge, causes of, 131, 133
  splash system, 124, 125, 133
  storage of, 134, 135
  tests, 128
  viscosity, 129

Maintenance, 186–191, 273
Management of plant, 292–294
  efficient operation, 292
  log sheets, 292, 293
  shift schedule, 294
Matter, three states of, 4
Meat-packing:
  modern, 201
  refrigeration, 201
  temperature range, 202
Mechanical refrigeration (*see*
  Refrigeration)
Mechanical shaft seal, 44, 138
Melting point, 1
Mercury, 7
Methane, 288
Moisture in system, 29, 188
Molecules, 6
Motors, electric, 146–158
  direct-current, 153
  enclosure types: dripproof, 154
    explosion proof, 166
    hermetically sealed, 150
    pipe-ventilated, 154
    splashproof, 154
    totally enclosed, 155
    weather-protected, 154

Motors, electric (*Cont.*):
  polyphase, 146, 147, 152
  rotation, changing direction, 155
  safety controls, 157
  single-phase, 146, 147, 153
    starting methods for, 149, 150
  slip-ring or wound-rotor, 149
  squirrel-cage: induction, 147
    starting method for, 148
  synchronous, 151

National Electric Codes (NEC), 155
National Electric Manufacturers
  Association (NEMA), 154
Noncondensables, 63

Odor, 239, 240
Oil lantern rings, 43, 139
Operation, 175–197
  absorption or steam-jet system, 182
  checking new system, 182
  keeping load balanced, 181
  low suction pressure, 182
  low temperatures, 186
  purging centrifugal system, 182
  shutting down ammonia system, 182
  surging in centrifugal system, 182
  what temperatures indicate, 175–177
Oxygen, 281

Packing, mechanical, 136–139
People load, effect of, on air
  conditioning, 227
Perishable commodities, 161
Piping and fittings, 85, 107-123
  backpressure regulator, 111
  bypass, evaporator-regulator, 110,
    113
  checking needed, 284
  discharge line, 109, 110
  flooded-coil system, 112
  halocarbon, 108
  liquid receiver, 108
  materials for refrigeration, 107, 115
  for parallel operation, 111
  in R-12 system, 107
  for room cooler, 110
  shell-and-tube chiller, 112, 113
  solenoid valve, 114
  suction for evaporators, 110

Piston rings, 189
Plate coil, 68
Pounds per square inch (psi), 7
Precooling fruit, 203
Pressure:
  absolute (psia, psig), 7
  critical, 8
  discharge, 177
  effect on boiling point, 6
  gage, 7
  high-side, 9
  low-side, 9
  partial, law of, 7
  suction, 9
  temperature, 8
Pressure drop:
  compressor to condenser, 107
  receiver to expansion valve, 107
  suction line, evaporator to
      compressor, 107
Psychrometric chart, 229–231
  finding humidity, 231
  wet-bulb readings, 231
Pump, centrifugal, 248
Pump sizing for cooling towers, 222
Pumping head, 209
Purge-recovery unit, starting, 252
Purging, 183
  automatic, 183
  centrifugal system, 182, 183
  Freon-12 system, 184
  valve, 83

Reaumur temperature, 1, 2
Receiver, liquid, connections for, 108
Refrigerant controls, 72–95
  for test panel, 92–94
Refrigerants, 27–36
  for chemical process, 174
  for construction, 173
  critical temperature, 28
  cylinders, handling, 285
  dangers of, 277
  misting in system, 35
  mixing with oil, 281
  moisture, effect of, 35
  numbering system, 31–34
  odorless, disadvantage of, 28
  physical properties of, 29, 30
  stability of, 29
  toxicity, 275

Refrigerants (*Cont.*):
  in vacuum, 29
  vapor density, 28
  weight, change in, 177
Refrigeration, 1
  applications of, 198–206
  five methods of, 14
  hazards of, 272
  ton of, 8
Refrigeration systems, 14–26
  absorption, 14
  air in, 181
  balancing load, 181
  booster, 23, 24
  centrifugal, 25
  checking new, 283, 286
  compression, 14–17
  management of, 292–294
  mechanical, principles of, 14
  steam-jet, 22
  temperatures in, 16
  thermoelectric, 14
  two-stage, 52
  vacuum, 22
Regulators, evaporator pressure, 87, 88
  air-modulated, 89
  pilot-operated, 87
Relative humidity, 228
Reverse operation, air conditioning,
    245
Rotary screw compressor (*see*
    Compressors)

Safety, 272–286
Safety controls, 103
Safety valves (*see* Valves, safety)
Salt, 14
Sanitary regulations, 242
Saturation temperature, 228
Shaft seals, 136
Shift schedule, 294
Spray cooler, 232
Spray humidifying, 240
Spray pond, 210
Sprayed roof, 246
Starting bypass, 48
Steam-jet system, 22
Storage, controlled atmosphere for, 203
Strainer, 81
Stuffing box, 43, 136
Suction-line regulators, 87
Superheat, 9, 176

Switches, electric, 176
  DPST, SPDT, SPST, 99
  float, 83–85, 102
  low oil-pressure, 100
Systems, refrigeration (*see*
  Refrigeration systems)

Temperature, 1
  control of, 98
  for efficient freezing, 200
  for frozen foods, 201
Temperatures for air conditioning, 247
Terminology, refrigeration, 10–12
Testing:
  of controls, 92–94
  for leaks, 186, 187
  safety valves, 188
Thermoelectric systems, 14
Thermometers, 2, 175
  in discharge line, 177
  in suction line, 176
Thermostatic expansion valve, 73
Thermostats, 97–99
  adjustable differential, 98
  bimetal, 98
  checking, 284
  cross-ambient fill, 99
  high-temperature, fade-out, or
    limited fill, 99
  locating, 251
  remote-bulb, 99
  turning back (at night), 251
Total heat, 229
Trane chart, 229
  air-conditioning cycle, 232
  air mixtures, finding, 229
  air volume, 232
Troubleshooter's guide:
  air conditioning, calculations for,
    247–251
  to electrical problems, 103–105
  to refrigeration problems, 191–197

Unloader, 47

V-belts, 156
Vacuum, 7, 22
Valves:
  crossover, 47, 142
  discharge, 45, 48, 190
  evaporator-pressure, 87
  expansion, 228
    checking, 284
    thermostatic, 116
  float, 83
  purge, 83
  safety: purpose of, 116
    regulating, 284
    testing of, 188
  snifter, 48
  solenoid, 85, 86, 91, 92
    direct-acting, 85
    liquid line, 90, 91
    pilot-operated, 85
    where used, 116
  suction, 90, 116
  water-regulating, 116
Vapor barrier, 121
Vent tube, 83
Ventilation, machine room, 282
Vibration in ducts, 246

Water conservation, 243
Water coolers, 71
Water jacket, 44
Water tower, securing, 258
Water vapor, superheated, 228
Wet- and dry-bulb temperatures, 208
Wet-bulb temperature, 208, 209, 229
Wind chill factor, 10

Zoning, 245